Ecological Studies

Analysis and Synthesis

Volume 46

William M. Lewis, Jr.
James F. Saunders, III
David W. Crumpacker, Sr.
Charles M. Brendecke

Eutrophication and Land Use

Lake Dillon, Colorado

With 67 Figures

Springer-Verlag
New York Berlin Heidelberg Tokyo

William M. Lewis, Jr.
Department of Environmental,
Population, and Organismic Biology
University of Colorado
Boulder, Colorado 80309
U.S.A.

James F. Saunders, III
Department of Environmental,
Population, and Organismic Biology
University of Colorado
Boulder, Colorado 80309
U.S.A.

David W. Crumpacker, Sr.
Department of Environmental,
Population, and Organismic Biology
University of Colorado
Boulder, Colorado 80309
U.S.A.

Charles M. Brendecke
Department of Civil and
Environmental Engineering
University of Colorado
Boulder, Colorado 80309
U.S.A.

Library of Congress Cataloging in Publication Data
Main entry under title:
Eutrophication and land use, Lake Dillon, Colorado.
(Ecological studies series ; v. 46)
Bibliography: p.
Includes index.
1. Eutrophication—Colorado—Dillon Reservoir.
2. Land use—Colorado—Dillon Reservoir Watershed.
3. Dillon Reservoir (Colo.) I. Lewis, William M., Jr., (date).
II. Series: Ecological studies ; v. 46.
QH105.C6F93 1984 574.5′26322′0978845 84-1245

Typeset by MS Associates, Champaign, Illinois.
Printed and bound by Halliday Lithograph, West Hanover, Massachusetts.
Printed in the United States of America.

9 8 7 6 5 4 3 2 1

ISBN 0-387-90961-3 Springer-Verlag New York Berlin Heidelberg Tokyo
ISBN 3-540-90961-3 Springer-Verlag Berlin Heidelberg New York Tokyo

Preface

Nutrient enrichment (eutrophication) is a major theme in freshwater ecology. Some themes come and go, but the inevitable release of phosphorus and nitrogen that accompanies human presence seems to ensure that eutrophication will not soon become an outmoded subject of study. Eutrophication raises issues that range from the pressingly practical problems of phosphorus removal to the very fundamental ecological questions surrounding biological community regulation by resource supply. Although it is possible to take a reductionist approach to some aspects of eutrophication, the study of eutrophication is fundamentally a branch of ecosystem ecology. To understand eutrophication in a given setting, one is inevitably forced to consider physical, chemical, and biological phenomena together. Thus while eutrophication is the focus of our study of Lake Dillon, we have assumed that a broad base of limnological information is a prerequisite foundation.

Eutrophication of a lake can be studied strictly from a limnological perspective. If so, the nutrient income of the lake is quantified but the sources are combined within a black box whose only important feature is total loading. It is also possible, however, to treat the watershed and lake as equally important components of a hybrid system. In this case the nutrient sources must be dissected and their variability and dependence on key factors such as runoff must be quantified. In view of the abundant information on nutrient flux of watershed, it seems increasingly difficult to justify continued separate treatment of lakes and watersheds. In the Lake Dillon study, we have attempted to treat watershed analysis as comprehensively as we have limnological analysis, and thus to achieve some insight into land-water coupling.

As a subject of case history analysis, Lake Dillon has some especially interesting features. Many eutrophication studies deal with grossly polluted lakes. Lake Dillon, which would be unproductive (oligotrophic) if the watershed were uninhabited, was brought to a moderately enriched (mesotrophic) status in the 1970s by watershed development. Development was not accompanied by unplanned disposal of minimally treated waste, but rather by a wasteload allocation sufficiently stringent to require tertiary treatment for phosphorus at all four major plants within the watershed. Thus Lake Dillon is an example of eutrophication in the face of state-of-the-art point source technology. Because point sources are subject to tight control, a wide variety of dispersed sources and natural background account for major proportions of lake nutrient loading. Lake Dillon is also of special interest in that it receives extraordinarily high ratios of nitrogen to phosphorus in runoff, yet experiences an interval of pronounced nitrogen limitation in the mixed layer. This seems inconsistent with current theory, and is thus a major focus of attention in our analysis.

The Lake Dillon study was supported by funds from the United States Environmental Protection Agency under a Clean Lakes Grant (Section 314, Clean Water Act) and by the Northwest Colorado Council of Governments, from the towns of Breckenridge, Frisco, and Dillon; sanitation districts of Breckenridge, Copper Mt., Frisco, and Dillon-Silverthorne; the Denver Water Board, and the AMAX and Keystone Corporations. The project was administered by the Dillon Clean Lakes Steering Committee, whose members and their constituencies have been exceptionally supportive and helpful. We are especially indebted to the Committee's secretary, Mr. Tom Elmore, who has answered innumerable queries and solved many problems for us, and to the chairman of the Committee, Mr. Bruce Baumgartner, for his decisive leadership.

Analytical work was done by the University of Colorado Limnology Laboratory. Chemists for the project included Ms. Katherine Ochsner, Mr. Stephen Hamilton, Ms. Margaret Robbins, and Mr. Lewis Dennis, all of whom were dedicated to their work well beyond what we had a right to expect. The field crew, which performed its duties well many times under the most arduous conditions, included at various times Dr. Robert Epp, Mr. Steven Murray, Mr. Donald Morris, Jr., Mr. George Kling, and occasionally our chemists and Mr. Terry Carter of the Colorado Department of Health, who was also helpful in many other ways. Ms. Mary Marcotte did a skillful job of preparing the manuscript and Mr. George Kling did the same for the drafting. We are grateful to Ms. Heidi Thompson for counting the zooplankton and to Dr. R. Dufford for confirming the diatom identifications.

We thank the Denver Water Department, Ms. Jerry Vest and other members of the Summit County Planning Department, the Summit County Engineering Department, Mr. Barry Sheakley of the U.S. Forest Service's Dillon Ranger District, and Mr. Wesley Nelson of the Colorado Division of Wildlife for supplying us with various kinds of data. We are grateful to the wastewater treatment plant operators and especially to Mr. Buck Wenger of the Snake River Wastewater Treatment Plant, where our field crews did a considerable amount of sample processing. We also appreciate occasional volunteer contributions from Dr. Thomas Frost, Dr. Suzanne Levine, and Ms. Claudia Cressa, all of the University of Colorado Limnology Laboratory.

Contents

1. Introduction

Lake Dillon of Summit County, Colorado, is an impoundment of the Blue River just below its confluence with the Snake River and Tenmile Creek. The watershed drains elevations between lake level at about 2750 m and the mountainous headwaters of the three inflowing rivers at elevations as high as 4300 m. For at least the first decade after its creation in 1963, Lake Dillon was considered unequivocally oligotrophic, as shown by its high transparency. Because of its location, Lake Dillon would be expected to remain oligotrophic indefinitely if the watershed were uninhabited or very sparsely inhabited. Watersheds at high elevations in the Central Rockies are seldom rich in phosphorus because the parent material, which lies relatively close to the surface, consists mostly of hard crystalline rock that is resistant to weathering and poor in phosphorus. In addition, the natural vegetative cover effectively holds the particulate phosphorus inventory, as shown by the clarity of streams in undisturbed areas, even at times of peak runoff.

The water of Lake Dillon is under the direct control of the Denver Water Department, which uses the lake as the main storage facility for the city of Denver. Recreational activities on the lake are managed by the U.S. Forest Service. The lake is a popular sport fishery for rainbow trout, brown trout, and kokanee salmon, and, because of a significant spawning run of brown trout up the Blue River, is of use to the state as a source of brown trout eggs. Numerous permanent homes and vacation homes lie within sight of the lake and most of the owners of these properties would probably regard the blue color and high transparency of Lake Dillon as an aesthetically or economically valuable amenity. Furthermore, the appearance of the lake is especially prominent to the general public because of the proximity of the lake to Interstate 70. Thus it is clear that major changes in the algal standing crop or transparency of Lake Dillon could be undesirable from diverse viewpoints, ranging from increased water treatment costs for the city of Denver and lower property values for lakeside residents to loss of aesthetic appeal for watershed residents and the general public.

The uses of the Lake Dillon Watershed are especially diverse. The U.S. Forest Service controls over half the land of the Lake Dillon watershed; its holdings are principally on the steeper slopes and at the higher elevations and are essentially undeveloped. At lower elevations, the watershed contains three major municipalities (Frisco, Breckenridge, and Dillon: Figure 1). There are many other smaller housing developments, some of which are served by small package plants, some by sewer through the four major treatment plants, and others by septic systems. Four important ski areas are also located in the watershed (Breckenridge, Copper Mountain, Keystone, and Arapahoe Basin). Climax Molybdenum, a major mining enterprise, is located at the headwaters of Tenmile Creek. Finally, the watershed serves a large number of nonresident visitors, who come in greatest numbers for winter sports and in somewhat smaller numbers for summer recreation. In 1981-1982, the time-weighted average population of the watershed was 19,000 persons and the seasonal peak (winter) was about 84,000 persons. According to the Summit County Planning Department, Sum-

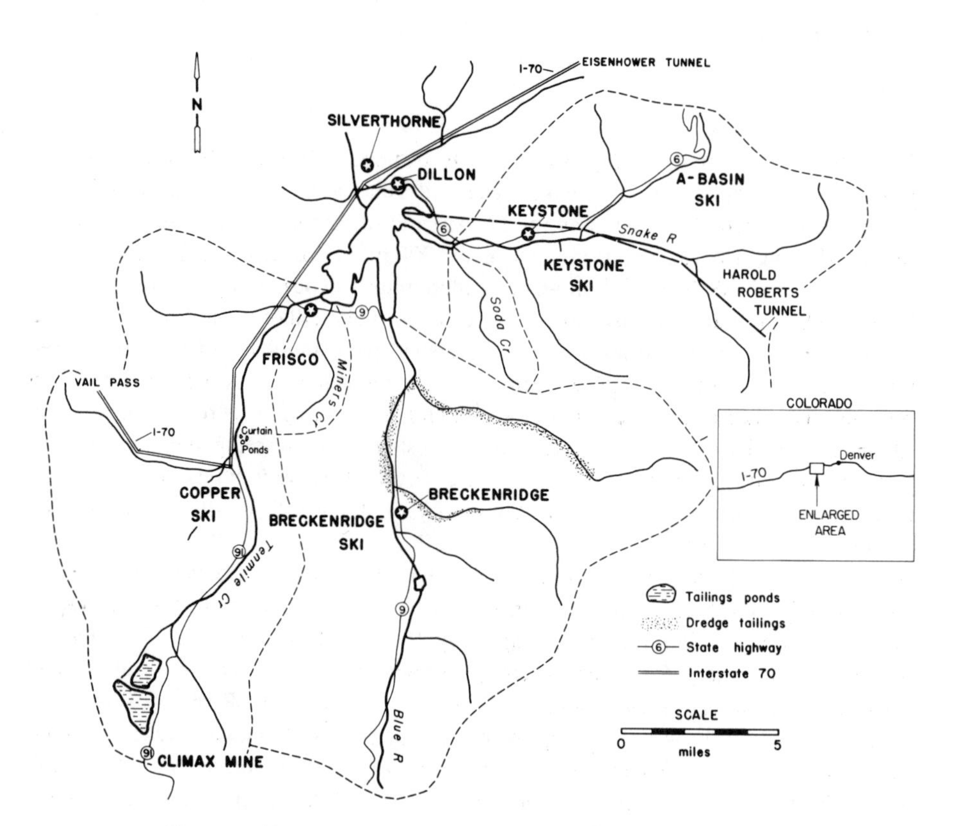

Figure 1. Orientation map of Lake Dillon and its watershed.

mit County, half of which is comprised of the Lake Dillon catchment, was the fastest growing county in the United States in 1981.

The background total phosphorus concentrations for the Dillon watershed in complete absence of human activity would be only about 5 μg/liter P. Such low background phosphorus concentrations can maintain lakes of very low algal standing crop and thus of high transparency and beautifully blue water. In contrast to the many lakes that occupy watersheds whose background total phosphorus concentrations are several times higher than this, relatively little human presence or human activity is required to double or triple the total phosphorus concentrations. The vast majority of lakes that have been studied with respect to eutrophication have higher background phosphorus concentrations than Lake Dillon, so good analogies are not always available. We will draw on information from the lakes of the Canadian Shield and parts of Scandinavia, since these lakes typically have low phosphorus concentrations, cool water, and short growing seasons reminiscent of Lake Dillon. Reservoirs and natural lakes in the midwestern and southeastern United States, while thoroughly studied, are less similar to Dillon in these important respects. Lake Tahoe comes to mind as a good comparison for Dillon and will occasionally be useful, but Tahoe is considerably more oligotrophic than Lake Dillon, owing largely to its great mean depth and small ratio of watershed area to lake volume.

The Lake Dillon study is based on field work extending from 1 January 1981 to 31 December 1982. In the following chapters, we first deal with the study design and methods of data collection. A comprehensive limnological analysis follows; the emphasis of this is on trophic status and nutrient chemistry. Next is an analysis of watershed nutrient yield, including quantification of total nutrient loading and separation of nutrient sources based on land use analysis. The last chapters are devoted to description and application of a model that integrates and applies the information on land use and trophic status.

2. Design of the Study

Variation in Lake Dillon and its watershed was quantified by a dual sampling program, one part of which was designed primarily to provide detailed information for a few sites on many dates and the second to provide a detailed picture of spatial variation on a smaller number of dates. These two approaches were supplemented by special data collection programs of more limited scope whose purpose was to provide information that would not necessarily be forthcoming from the routine studies. The Lake Dillon Study is thus supported by three data sets: (1) time series, (2) spatial survey, and (3) special studies.

Time Series Data Set

The time series data consist of routine lake and stream data and precipitation chemistry data.

Routine Lake Data

A key station was selected near the center of the lake in deep water (Figure 2). On each routine lake sampling date, the water column at this "index station" was sampled in 5-m increments from top to bottom. Variables quantified there are listed in Table 1. In addition, one "main" station was selected at the mouth of each of the four major arms of the lake (Figure 2). The main stations were always sampled as part of the rou-

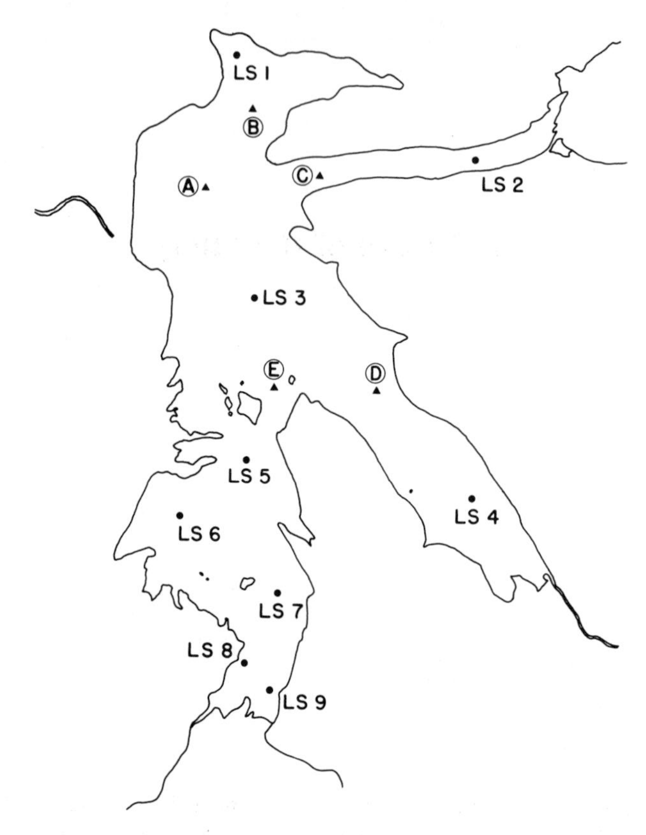

Figure 2. Map of Lake Dillon showing the locations of sampling stations, including the index station (A), the four main stations (B–E), and the survey stations (LS 1–LS 9).

tine lake sampling program. However, samples were taken only at the top and bottom of the water column rather than over the entire vertical profile as at the index station. Analytical coverage is summarized in Table 1. The schedule of sampling for the routine lake data is shown in Table 2. Samples were taken on 32 different dates. For most of the year, the collections were biweekly, but during the period of ice cover collections were less frequent.

Routine Stream Data

All overland flow was sampled as close to the lake as possible. This required collections at eight stations; the resulting data make up the routine stream data set. The eight stations, with abbreviations as shown in Figure 3, were as follows: Snake River Mouth (SR1), Blue River Mouth (BR1), Tenmile Creek Mouth (TC1), Miner's Creek Mouth (MC2), Soda Creek Mouth (SC2), Blue River Outlet (BRO), Frisco Effluent, and Snake River Effluent. Other effluents were represented in the river mouth samplings because they do not enter the lake directly. The analytical coverage and sam-

Table 1. Summary of Analytical Coverage for the Different Kinds of Samples

Variable	Time Series				Spatial Survey	
	Lake Index Station	Lake Main Station	Routine Stream	Precipitation	Lake Survey	Stream Survey
Temperature	+	+	+			+
Conductance	+	+	+	+	+	+
Transparency	+	+			+	
Discharge			+			+
Dissolved O_2	+	+				
pH	+	+	+	+		
HCO_3^-	+	+	+	+		
Soluble reactive P	+	+	+	+	+	+
NH_4^+-H	+	+	+	+	+	+
NO_2^--N	+	+	+	+	+	+
NO_3-N	+	+	+	+	+	+
Total soluble P	+	+	+	+	+	+
Total soluble N	+	+	+	+	+	+
Particulates	+	+	+	+	+	+
Partic. C	+		+			
Partic. P	+	+	+	+	+	+
Partic. N	+		+			
Chlorophyll *a*	+	+			+	
Primary production	+					
Phytoplankton	+					

Table 2. Summary of Sampling Dates[a]

Routine Lake/Stream	Stream Survey	Lake Survey	Diel	Enrichment
		1981		
23 Feb	26 Jan	09 Feb	14 Jan	20 July
30 March	27 Apr	18 May	16 Nov	24 Aug
13 Apr	05 May	29 June		28 Sept
11 May	18 May	20 July		
26 May	01 June	24 Aug		
08 June	08 June			
22 June	15 June			
13 July	06 July			
27 July	03 Aug			
10 Aug	05 Oct			
31 Aug				
14 Sept				
28 Sept				
12 Oct				
26 Oct				
09 Nov				
23 Nov				
21 Dec[b]				

Table 2. *(Continued)*

Routine Lake/Stream	Stream Survey	Lake Survey	Diel	Enrichment
		1982		
01 Mar	25 Jan	08 Feb		24 May
15 Mar	10 May	24 May		14 June
05 Apr[b]	31 May	14 June		10 July
19 Apr[b]	07 June	26 July		26 July
03 May[b]	14 June	23 Aug		23 Aug
17 May[b]	21 June			07 Sept
07 June	28 June			18 Oct
21 June	12 July			
06 July	09 Aug			
19 July	11 Oct			
02 Aug				
16 Aug				
07 Sept				
20 Sept				
04 Oct				
18 Oct				
01 Nov				
15 Nov				
06 Dec[b]				

[a] Precipitation chemistry samples were collected on the same schedule as routine lake and stream samples.
[b] Streams only; lake work unsafe.

pling schedule are shown in Tables 1 and 2. Discharge was also estimated at each sampling station on each date.

Precipitation Chemistry

Bulk precipitation (the combination of dryfall and wet precipitation) was sampled continuously near the Snake River Wastewater Treatment Plant (Figure 3). The sample was collected whenever samples of any other kind were collected on the lake or in the watershed. The analytical coverage is summarized in Table 1.

Spatial Survey Series

The spatial survey series consists of two parts: lake survey and stream survey.

Lake Survey

The lake survey stations include the index station, the four main stations, and nine additional stations scattered over the lake (Figure 2). Analysis for each station was based on an integrated sample taken from the top 5 m. A bottom sample was also taken at the index station and four main stations. The analytical coverage is summarized in Table 1. The survey was repeated 10 times over the course of the 2-year study

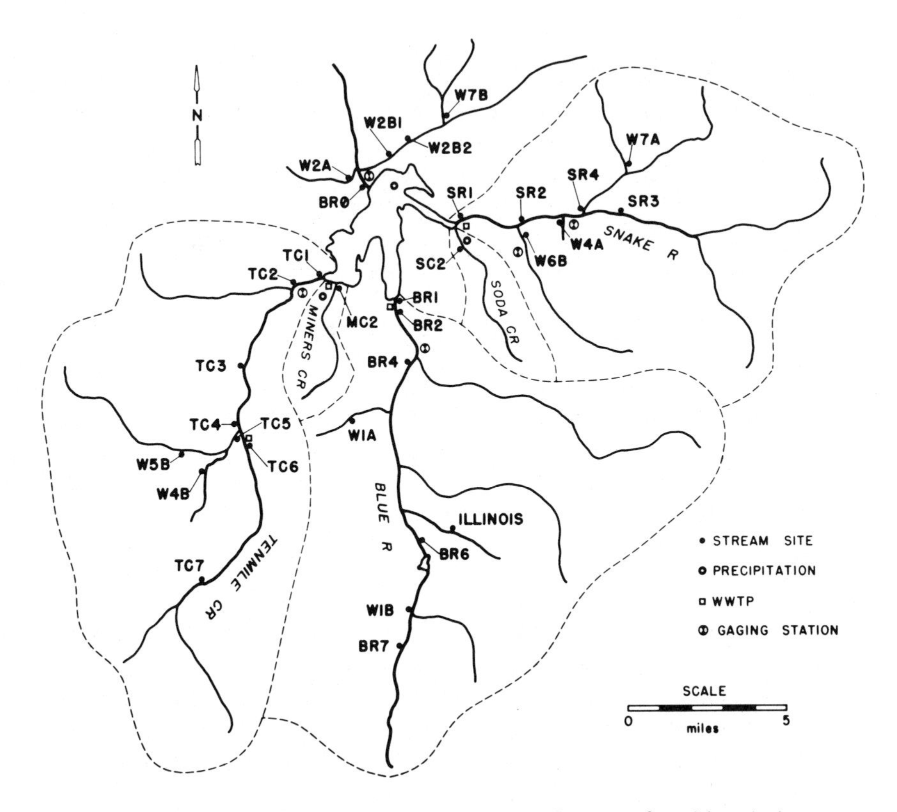

Figure 3. Locations of stream sampling sites. See text for abbreviations.

interval, thus allowing for seasonal changes in spatial heterogeneity over the surface of the lake.

Stream Survey

The stream survey series is based on samples taken at 33 sites in the watershed as shown in Figure 3. As with the lake survey, all stations were sampled on the same day so that comparisons could be made between stations for a given date. The 33 stations include the stream survey stations already mentioned above and the 11 stations of the representative watershed sampling program (to be discussed below), plus 14 other stations distributed up and down the main stems of the three rivers flowing into Lake Dillon. The sampling dates are shown in Table 2. A few of the stations were not sampled over the entire 2 years; these will be identified in the course of data analysis.

Special Studies

Enrichment Studies

On 10 different dates, water samples from the upper water column were enclosed in flexible containers and suspended within the lighted zone of the lake. Inorganic phosphorus was added to certain of these containers, inorganic nitrogen to others, and a combination of nitrogen and phosphorus to still others. Another set of containers was left unaltered as a control. Growth responses to the various treatments were determined by changes in the concentration of chlorophyll *a* within the containers. The objective of this work was to determine whether the phytoplankton community at a given time was limited by phosphorus, by nitrogen, or by neither of these elements. The enrichment experiments were done more frequently in 1982 than in 1981 after preliminary evidence indicated a late-summer switch between phosphorus and nitrogen limitation. The schedule of enrichment experiments is given in Table 2.

Diel Studies of Nutrient Chemistry

On two occasions the nutrient transport to the mouth of the Blue River, the Snake River, and Tenmile Creek was studied over a 24-h period. The objective was to determine whether or not significant bias would result if loading of the lake with phosphorus and nitrogen were estimated on the basis of samples taken at a particular time of day or exclusively during the daylight hours.

Sediment Chemistry

Lake sediment samples were taken at the four main stations and at the index station on 26 July 1982. The interstitial waters and particulate component of the sediments were analyzed separately for all phorphorus fractions. The objective of this work was to show whether or not substantial reserves of phosphorus are present in the sediments of Lake Dillon and, if so, whether or not significant amounts of these reserves are present as soluble phosphorus in the interstitial waters.

Representative Watershed Program

Complete analysis of present phosphorus sources and modelling of future phosphorus loading both require knowledge of nutrient export coefficients associated with different land uses in the Lake Dillon watershed. Export coefficients cannot be derived from data on high-order streams because land uses are typically mixed in the watersheds of such streams. For this reason, a set of representative watersheds was chosen, each of which is drained by a low-order stream and whose land use is readily assignable to a particular unmixed category. The land-use categories chosen for representation are as follows: undisturbed forested watershed, residential areas served by septic systems, residential areas served by sewer systems, urban areas served by sewers, ski slopes, roads, mining, and interstate highway. Two watersheds were chosen for most types. In some instances, only one watershed for a particular type could be found, and some of the sites were moved after the first year. The discharges and concentrations of nutrients for each of the watersheds were determined for each stream survey sampling date. These data were used to compute the nutrient output per unit area for each of the watersheds.

Supplemental Precipitation Chemistry

After the 1981 data showed that precipitation makes a significant contribution to the total phosphorus loading of Lake Dillon, it was decided that data on atmospheric phosphorus deposition should be taken at several sites. In the summer of 1982, one additional sampler was set up near the Frisco Wastewater Treatment Plant and another was placed on the Denver Water Department's raft on the lake surface. The purpose of the latter collector was to show whether the collection stations on shore were unduly influenced by terrestrial dusts.

Groundwater Study

Water and nutrients enter Lake Dillon from the air, as overland runoff, and through groundwater. In view of the geology of the Lake Dillon watershed, groundwater movement was considered to be a minor source of nutrients for the lake, but a study was nevertheless made of the groundwater flow in order to verify this impression.

Land Use Study

Partitioning of the present nutrient loading of the lake according to source requires quantitive information on land use. Our land use study of the Lake Dillon watershed provided estimates of the intensity, distribution, and total amount of various land uses.

3. Methods

Lake and Stream Sample Collection

Lakewater samples were always taken with an integrating sampler of the type described by Lewis and Saunders (1979). The sampler consists of a PVC tube 5 m long with closure devices at each end that can be triggered by a messenger. When the sampler is retrieved to the boat, the water is released through a piece of surgical rubber tubing into an integrating chamber, where the contents of the sampler are thoroughly mixed. Water is then drawn from the integrating chamber into sample bottles. The integrating chamber incorporates a baffle that prevents contact between the air over the sample and the water, thus allowing collection of samples for analysis of oxygen or other atmosphere-sensitive variables.

The advantage of the integrating sampler is that it eliminates sampling variance associated with vertical layering. When successive 5-m increments are taken from the top to the bottom of a water column, no layers will have been missed. At the same time, since the sampler can be used on successive 5-m increments in the water column, information on gross vertical structure of the water column can be obtained with the sampler.

Samples of Lake Dillon were protected from heat and from direct exposure to sunlight after they were bottled in the field. Special precautions were taken to protect the chlorophyll samples from exposure to light. Sample bottles were always rinsed copiously with deionized water and dried at the end of each sampling period.

Streamwater samples were collected with a wide-mouth plastic bottle held just below the water surface in a reach of the stream where the flow was fast across the entire cross section. Care was taken not to stir up stream sediments in front of the sample.

Filtration and Analysis of Particulates

Filtration of water samples was carried out at the Snake River Wastewater Treatment Plant (Figure 3) immediately after sample collection. Each sample was shaken prior to filtration and poured onto a filter tower over a Whatman GF/C glass fiber paper (47 mm diameter, pore size ~2 μm). The purpose of the immediate filtration was to reduce the biological activity of the sample to a minimal level without the addition of preservatives that might affect sample chemistry. The filtered samples were then returned to Boulder where they were stored under refrigeration prior to analysis. Analysis of soluble constituents began the morning after sample collection. Analysis of labile constituents was completed the day after sampling.

One filter was used for the analysis of particulate phosphorus (see below), and, if chlorophyll was to be measured, a second filter was used for this. The amount of water that would pass through a single filter depended on the amount of particulates in the sample, but typically fell between 100 and 1000 ml.

Another filter for each sample was dried to constant weight at 60°C and weighed with an analytical balance. A measured amount of sample water was poured over this filter, after which the filter was redried to constant weight at 60°C in an oven, cooled in a desiccator to room temperature, and weighed to the nearest 0.01 mg on an analytical balance. The difference between the initial and final weights was recorded as the amount of particulate material on the filter, which was subsequently converted to milligrams per liter of total particulates. The filter was then analyzed for carbon and nitrogen with a Carlo Erba Model 1102 elemental analyzer. Because the amount of nitrogen in the particulate fraction was a very small porportion of the total nitrogen, especially for the streams, the elemental analysis was done only on the samples from the river mouths, the lake index station, and selected additional samples from various locations.

Phosphorus Analysis

One filter from each sample was analyzed for particulate phosphorus following the method of Solórzano and Sharp (1980a), which relies on decomposition of organic phosphorus compounds by pyrolysis. The filters are dried with magnesium sulfate, heated at 450-500°C for 1-2 hr, treated with hydrochloric acid to hydrolyze polyphosphates, and then analyzed for orthophosphate.

Soluble reactive phosphorus was measured by the ascorbic acid-molybdate method (Murphy and Riley 1962). A 10-cm cell was used for low levels of phosphorus. This allowed measurement of as little as 1 μg/liter of P. Subtraction of a turbidity blank at 885 nm is sometimes necessary but proved of negligible importance on the filtered samples of the Dillon study. Although the chemical species measured by this test is sometimes referred to as PO_4-P, it is more properly characterized as soluble reactive phosphorus (SRP), since the test is sensitive not only to orthophosphate but also to organic phosphorus compounds of low molecular weight (Rigler 1968, Levine and Schindler 1980).

A pyrolysis method was used for total soluble phosphorus (Solórzano and Sharp 1980a). A portion of the filtered sample is evaporated to dryness with magnesium sulfate. The sample is then combusted at 450-500°C and the resulting polyphosphates are hydrolyzed with hydrochloric acid. This is followed by determination of soluble reactive phosphorus by the molybdate method. Subtraction of the soluble reactive phosphorus from total soluble phosphorus gives soluble organic phosphorus (or, more properly, nonreactive soluble phosphorus). Addition of total soluble phosphorus to particulate phosphorus gives total phosphorus.

Soluble Nitrogen Analysis

A modification of Solórzano's phenolhypochlorite method (1969) was used for ammonium (see Grasshoff 1976). Hypochlorite is added to ammonium in alkaline solution. With the addition of phenol and nitroprusside, indophenol blue is formed. The indophenol blue concentration is determined spectrophotometrically with a cell up to 10 cm in length as appropriate.

The widely used diazotization method originally described by Bendschneider and Robinson (1952) was used for nitrite. The combination of nitrate plus nitrite was determined by passage of a portion of filtered sample through a reduction column containing a cadmium-copper couple, after addition of a buffer (Wood et al. 1967). Corrections were routinely made for column efficiency. After May 1982, all nitrate measurements were made by ion chromatography rather than wet chemistry. Ion chromatography was done with a Dionex Model 2110 chromatograph and bicarbonate eluant. The ion chromatographic method is highly sensitive and has lower error variance because of reduced sample handling.

A new method outlined by Solórzano and Sharp (1980b) was used for total soluble nitrogen. Analysis of small amounts of soluble organic nitrogen has historically been unsatisfactory because of low sensitivity or poor recovery. The new method, which emphasizes better control over pH and oxidizing conditions, reduces these problems. The method is based on oxidation of soluble organic nitrogen by potassium persulfate in an autoclave. This converts all soluble nitrogen compounds to nitrate. Nitrate is subsequently analyzed by the reduction method described above. This produces an estimate of total soluble nitrogen. Soluble organic nitrogen can be obtained by subtraction of the ammonium, nitrate, and nitrite obtained in the other analyses. For the low levels of soluble organic nitrogen present in some samples of the Dillon study, recrystallization of reagents was necessary to improve the precision of the test.

pH, Alkalinity, Oxygen, and Major Ions

pH was determined with a Radiometer M29 pH meter and combination electrode on an unstirred sample. Alkalinity was determined by titration of a 100-ml aliquot to an endpoint of pH 4.4 with N/44 H_2SO_4. Dissolved oxygen was determined by the azide modification of the Winkler method (APHA 1975). Major ions not of direct nutritional significance were not determined routinely. A few samples were analyzed for major ions, and in these instances chloride and sulfate were determined by ion chromatography (Dionex 2110) and the cations were determined by atomic absorption spectrophotometry (Varian AA6) after addition of lanthanum oxide and cesium chloride.

Chlorophyll *a*

Chlorophyll *a* was determined after hot methanol extraction. The basic procedure is described by Talling (1974), but has been modified some in accordance with recommendations presented in the recent comprehensive methods evaluation edited by Rai (1980). The sample is filtered just after collection onto Whatman GF/C paper. $MgCO_3$ suspension (0.2 ml) is added to the filter, which is then placed in a screwcap test tube with 40 ml of 90% methanol, heated to 65°C, and allowed to boil gently for 30-60 sec. The tubes are then capped again and stored in the dark for 6 to 24 hr. The absorbance is measured at 665 and 750 nm relative to 90% methanol. The absorbance is then computed as follows: $12.9 \times (E_{665}-E_{750}) \times$ (volume of solvent in milliliters per liter filtered) $\times$ reciprocal of pathlength in centimeters. The coefficient of this equation is lower than Talling's original coefficient (13.9). The lower coefficient brings the calculation into line with recent information on the specific absorbance of chlorophyll in methanol (Marker et al. 1980). The data were also corrected for a problem that has not been brought out in the literature but which was discovered as a result of a seven-laboratory methods study coordinated by the University of Missouri (M. Knowlton and R. Jones, personal communication). Certain glass filters release a binder or fine colloidal fragment that does not centrifuge readily. These particles differentially increase the absorbance at 665 and 750 nm. Tests on our own filters showed that the fragments were present and increased the absorbance at 665 nm by 21.5% more than the absorbance at 750 nm. All of the calculations were corrected for this differential. The resulting difference in chlorophyll values was not very great in percentage terms, however.

The methanol method was selected because recent studies have shown its extraction capabilities to be better than those of any other method (Riemann 1980). We did a comparison of methanol and acetone specifically for Lake Dillon on 1 March 1982. Table 3 shows the results. As expected from the literature, methanol is more efficient

Table 3. Comparison of Extraction by Acetone and Methanol for Unmacerated Lake Dillon Samples on 1 March 1982

	Chlorophyll *a* (μg/l)		
Site	Methanol	Acetone	Ratio M/A
Tenmile arm	6.6	5.6	1.18
Index (0-5 m)	5.9	5.4	1.10
Index (5-10 m)	3.5	3.35	1.05
Index (10-15 m)	3.1	2.4	1.28

than acetone for samples that are steeped but not macerated in an homogenizer. The differential averages about 15%. According to the interlaboratory comparison, macerated samples treated with acetone would yield results very similar to those for samples steeped in hot methanol.

Phytoplankton and Zooplankton

Phytoplankton subsamples were taken of the samples for water chemistry from different depths. These were preserved in the field with Lugol's solution. Samples were subsequently counted by the method of Utermöhl (1958) with a Wild M-40 inverted microscope (1400X oil immersion). This method allows even the smallest cells to be distinguished and counted.

Zooplankton were sampled with a townet of 37-μm mesh. The net was lowered to 20 m and then raised to the surface. On the basis of general experience with metered nets of this mesh, we assumed that the efficiency of the net was 50%. The zooplankton were preserved in Formalin and counted by procedures similar to those described in Lewis (1979).

Primary Production

Primary production was determined by the C-14 uptake method at the index station on each routine lake sampling date. The method was carried out essentially as described by Lewis (1974). Depths of sampling and incubation for the Dillon work were 0, 0.5, 1, 2, 5, and 10 m. Greater depths are of little interest from the viewpoint of carbon fixation because there is insufficient light to support much photosynthesis. At each depth, two 125-ml transparent glass bottles were suspended ("light bottles"). One dark bottle was also filled with water from each depth. All of the dark bottles were incubated together at a depth of 5 m. Dark uptake of carbon was very small for Lake Dillon.

All bottles were inoculated with 1 ml of C-14 in the form of sodium bicarbonate solution (3 μCi/ml). The samples were suspended in situ for a timed period of 2.5-4 hr, typically between 10 A.M. and 2 P.M. On the day of the incubation, solar radiation reaching lake surface was recorded continuously by a Belfort pyrheliometer so that the amount of carbon fixation could be related to the amount of light.

At the end of the incubation, the bottles were removed from the lake, placed in a dark box, and transported to the Snake River Wastewater Treatment Plant for filtration. As each bottle was removed from the dark box, it was shaken and a 100-ml subsample was withdrawn for filtration. This subsample was placed onto a filter tower over a Millipore HA filter (pore size 0.45 μm). The sample was filtered under gentle vacuum and the filter was washed two times. The filter was then placed in a liquid scintillation vial containing 12 ml of scintillation cocktail (Aquasol II) and counted on a Beckman LS3133T scintillation counter. Counting efficiency was determined by addition of standard amounts of C-14-labeled toluene to selected samples. Counting efficiencies were very stable in the vicinity of 80-90%.

The total activity in each scintillation vial was computed from the observed activity and the machine efficiency after correction for machine blank (typically a very small correction in the vicinity of 50 cpm). The available inorganic carbon in the sample was determined from the pH and alkalinity of the sample and the sample temperature according to the tables of Saunders et al. (1962). Carbon fixation in each bottle was computed from the following information: (1) the activity of the sample, (2) the volume of water filtered, (3) the available carbon in the sample, (4) the isotope discrimination factor (1.06), and (5) the amount of C-14 added to the sample. The dark bottle fixation for a given depth was then subtracted from the light bottle fixation for the same depth to yield the observed photosynthetic carbon fixation for the incubation period (mg C/m^3/incubation period). This is considered to be a good estimate of gross photosynthesis (Harris 1978).

From the gross photosynthesis at each depth, a vertical profile could be made of photosynthesis at the index station over the incubation period. A profile of this type was drawn on a piece of graph paper for each one of the incubation dates. The amount of fixation over all depths was obtained by integration of the area under this volume-specific fixation curve. This yielded an estimate of gross fixation per unit area (mg C/m^2/incubation period). Because the incubation periods were of slightly different lengths and the sunlight conditions were different from one day to the next, fixation per unit area was converted to fixation per day and to fixation per unit of light. The total photosynthetically available radiation (PAR) striking the surface of the lake over the incubation period was obtained from the pyrheliometer sunlight trace by planimetric integration and computational adjustments as given in Chapter 8. The fixation per unit area for the incubation period was divided by the amount of incident PAR during the incubation period to yield the fixation rate/unit area/unit of PAR (mg C/m^2/ly). The PAR striking the surface of the lake over the entire day was then obtained by planimetric integration from the pyrheliometer charts. The total amount of PAR for the entire day was multiplied by the fixation per unit of light to obtain the total estimated fixation per unit area for the day (mg C/m^2/day). This last calculation assumes that, on a given day, the photosynthesis below a unit area is proportionate to the amount of PAR striking the surface. This assumption is not strictly correct (Talling 1971), but provides a close enough estimate for present purposes.

Recent studies have shown that some significant inaccuracies may result from confinement of samples at fixed depths because the transient and long-term response of algae to given light intensities are not always the same (Harris 1978, 1980). The magnitude of error that can result is not yet clear, but the data for Lake Dillon, as well as virtually all available field photosynthesis data, are subject to this error. The error will presumably affect absolute more than relative values.

Temperature, Conductance, and Transparency

Temperature profiles were measured with a YSI thermistor and submersible probe. Conductance was determined in the laboratory with a Labline Model MC-1 meter standardized against KCl. All conductances were temperature corrected to 25°C.

Transparency of the lake was determined by secchi disk, and sometimes also by submersible photometer (Whitney-Montedoro with submersible selenium photocell and opal filter). The photometer readings were taken from the surface to the limit of detection in increments of 60 cm. In Lake Dillon, the secchi depth was on the average equal to the depth of 19% surface irradiance as measured by light meter (standard deviation, 5.7%).

Discharge Measurements

Continuous discharge measurements were obtained from four USGS gauging stations. The USGS gauging station measurements serve two purposes: (1) they allow us to check our own methods of measuring discharge, and (2) they are the main basis for estimating total amount of water entering the lake from each of the three major rivers. The gauging station locations are shown in Figure 3.

The Snake River gauge is well above the mouth of the Snake River. In order to estimate the amount of water entering the lake from the Snake River, we added the values from the USGS gauges on the Snake River and on Keystone Gulch (a tributary of the Snake) and then made a small correction for the remaining ungauged portion of this drainage. Similar corrections were made for the other gauge readings to account for small areas below gauges (Figure 3). For the Blue River we also added the sewage treatment plant discharge to the gauge reading. The sewage treatment plant discharge was obtained from plant operators on sampling days.

We made discharge measurements when we took stream samples in ungauged reaches of stream for the routine stream series and for the stream survey series. Some of the stream sampling sites were sufficiently close to culverts or bridges that the discharge could be estimated by the use of standard tables (Chow 1959, BPR 1965). This was done for about eight sites in 1982 and slightly more in 1981. For most of the stream sites, standard tables could not be used because of the irregularity of the channel. Consequently, estimates were made by the method of cross-sectional current profiles. The depth profile of the stream channel was mapped at a location where the bed was likely to be stable. The cross-sectional area of the stream as shown by this map was divided into several segments (three to eight, depending on the width and irregularity of the stream). Within each segment, a current velocity measurement was taken with a Gurley Pygmy current meter at a depth equal to 60% of the distance from the surface to the stream bed. This is the depth at which the velocity is considered to be about average for a segment (Hynes 1970). The discharge of each segment was then computed from its cross-sectional area and its velocity of flow. The discharges for the different segments were added to produce the estimate of total discharge for the site. The current meter method is compared with the USGS data in Table 4. The deviations fall within the daily range of fluctuations for a given site and are thus as close as could be expected.

After a range of discharges had been obtained at a given site, a depth–discharge relationship was established in the form of a rating curve. When the rating curve was available, discharge estimates were sometimes based on the curve rather than current meter measurements.

The Denver Water Department provided daily estimates of water leaving the lake through the Roberts Tunnel and by the Blue River, and daily estimates of the total amount of water entering the lake.

Table 4. Comparison of Discharge Measurements Made by USGS and Near the Same Location by the Current Meter Method Used in the Field Sampling Program (1982)

Station	Number of Determinations	USGS (Mean, cfs)	Current Meter (Mean, cfs)	Diff. (%)
Tenmile Creek	20	223	216	−3
Blue River	21	173	167	−4
Snake River	21	145	131	−11

Precipitation Sampling

Atmospheric transport of nutrients was estimated near the Snake River Wastewater Treatment Plant by means of a large bulk precipitation chemistry sampler mounted on a tower 3 m above the ground. The design of the sampler is as given by Lewis and Grant (1978; Figure 4). The opening of the sampler is 0.20 m^2, which is sufficiently large to supply measurable amounts of phosphorus and nitrogen over weekly or biweekly collection periods at any time of the year, even when there is no wet precipitation. The collecting surface consists of an inverted Plexiglas pyramid that funnels materials into a narrow opening that leads to an insulated wooden box. Wet precipitation flows through the opening into a plastic receiving funnel. The receiving funnel is covered with a fine nylon screen so that insects reaching the funnel can crawl from it. The funnel is connected to a surgical rubber hose that is looped in order to form a vapor barrier, and the surgical rubber hose is connected to a hard glass receiving vessel with a collapsible plastic overflow container. The box housing the containers is heated by means of thermostatically controlled light bulbs. The heat is sufficient to melt snow that strikes the surface of the collector during the winter so that these contributions are not blown from the sampler.

The samples were collected at weekly to biweekly intervals depending on the schedule of other sampling programs. On each occasion, the collection vessels were removed from the box and mixed in a common container. An aliquot of 100 ml was removed for determination of pH and alkalinity. The surgical rubber hose was then hooked to a rinse container and 500 ml of deionized water was poured over the collecting surface, thus rinsing the collecting surface, collecting funnel, and tubing. After the volume of the rinse and the volume of wet precipitation were noted, the two were mixed together. The water was then analyzed according to the protocol for lake samples. The data were corrected for the addition of deionized water and for the removal of the 100-ml aliquot for pH and alkalinity determination. Concentrations were converted to loading rates on the basis of the area of the collecting surface and the elapsed time since the last sample.

Special Studies

The protocol for each of the special studies is described along with the results in later parts of this report. All chemical analyses and sampling were carried out by the methods described above unless otherwise indicated.

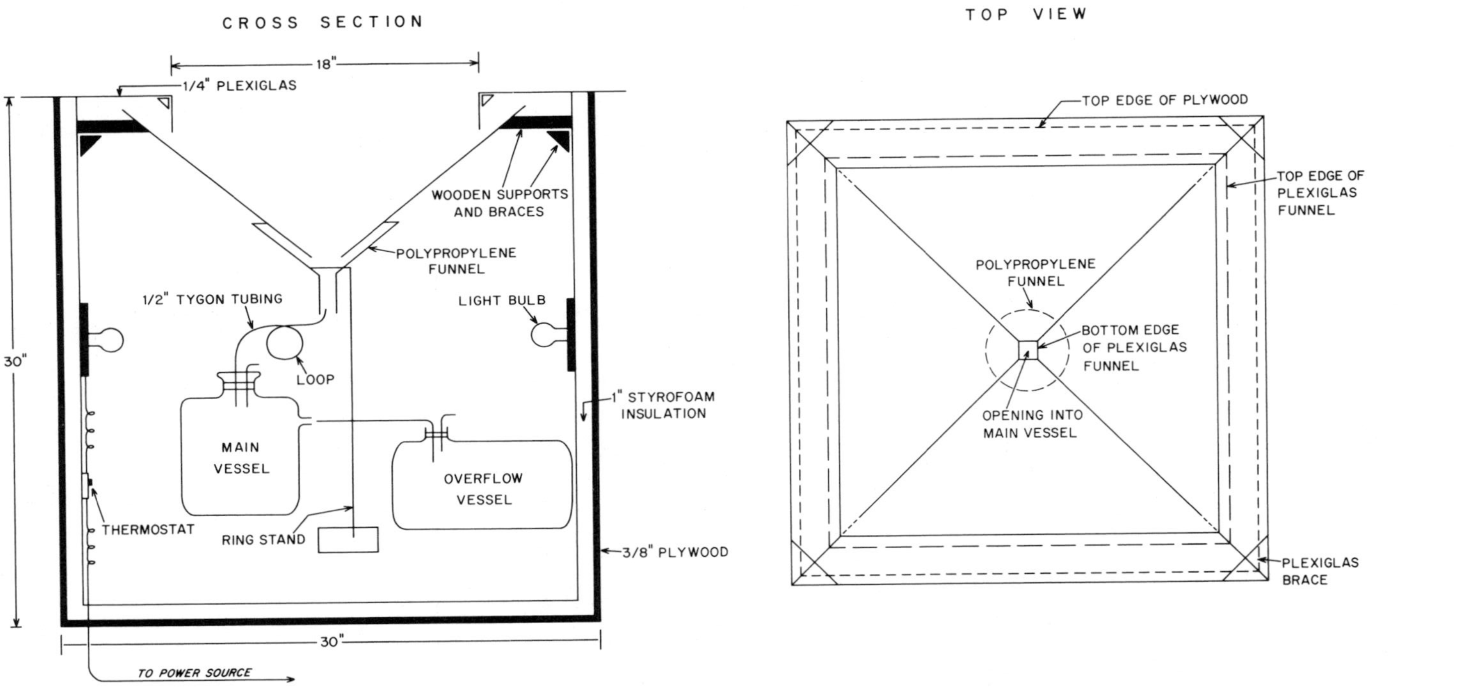

Figure 4. Design of precipitation chemistry sampler (from Lewis and Grant 1978).

Precision and Accuracy

Duplicates

Approximately 10% of the analyses were duplicated. Table 5 summarizes the resulting information. The degree of variability is most easily evaluated from the ratio of standard deviation to mean (coefficient of variation, expressed as percent), which is shown in the last column of Table 5. The degree of variation for all variables is acceptable, and for most it is very low and thus warrants no discussion. Estimates of total particulates have the highest relative variation. We attribute this largely to irregularities in the distribution of particles in streams (duplicates for stream analyses are from separate samples taken sequentially) rather than to variation in drying or weighing. This conclusion is supported by a difference in replicate variability for particulates in stream samples and lake samples. Lake duplicates, which come from the same integrator, and which are more likely to have an evenly dispersed particulate load of fine grain, show lower variability than stream duplicates. Even so, determination of the particulate load of streams will not be affected in any important way by this variability because of the large number of samples taken over time at each site. Replicate variability in particulate load is reflected to some degree in the particulate phosphorus analyses, as expected.

Soluble reactive P shows slightly higher relative variation of duplicates than most other constituents. This is explained by the low levels of SRP characteristic of the entire system (note that the average SRP concentration for all duplicates is just slightly over 3 μg/liter). Since the relative variation of a determination is to some degree affected by the proximity of the analysis to the analytical limit for the test, the slightly higher relative variation for SRP is not surprising.

Total soluble nitrogen also shows slightly higher variability than most other vari-

Table 5. Summary of Statistics on Duplicate Analyses

Variable	Number of Duplicates	Mean	Standard Deviation	Coefficient of Variation (%)
Oxygen	80	7.59 mg/l	0.095	1.3
Conductance	188	164.0 μmho/cm	4.4	2.7
pH	110	7.48	0.043	0.6
Alkalinity	115	36.8 mg/l	0.76	2.1
Chlorophyll *a*	101	7.29 μg/l	0.52	7.1
Total particulates	212	3.15 mg/l	0.56	17.8
NO_2-N	235	6.76 μg/l	0.37	5.5
NO_3-N	308	151.61 μg/l	5.91	3.9
NH_4-N	283	91.63 μg/l	5.59	6.1
Total sol. N	143	618.0 μg/l	78.0	12.6
Soluble reaction P	240	4.56 μg/l	0.40	8.8
Total sol. P	303	7.45 μg/l	0.56	7.5
Particulate P	224	4.95 μg/l	0.58	11.6

[a] Ratio of standard deviation to mean, times 100.

Table 6. Summary of Information on Spikes

Variable	Number of Spikes	Mean Percentage Recovery
Soluble react. P	213	91.4
Total sol. P	186	87.1
NO_2-N	187	104.7
NO_3-N	192	91.3
NH_4-N	196	96.7
Total sol. N	113	94.3

Table 7. Data on EPA Standards (Raw Standards Were Diluted to Appropriate Range)

	EPA	Detected Value
NH_4-N (μg/l)	19	19.3
NO_3-N (μg/l)	31	33.4
PO_4-P (μg/l)	3.1	3.0
pH	7.4	7.7

ables. We attribute this mostly to the pH sensitivity of the test, but do not consider it important in the interpretation of results.

Spikes

Table 6 summarizes the percentage recovery of spikes added to samples throughout the course of the study. The recovery is typically close to 100%. Only the total soluble phosphorus and nitrate recovery warrant discussion. For total soluble phosphorus, two types of spikes were used: dibasic inorganic phosphorus (K_2HPO_4) and ATP. The recovery of K_2HPO_4 was lower than for ATP, which was characteristically recovered near the 100% level. Since K_2HPO_4 is not expected in nature, whereas organic phosphorus compounds more similar to ATP are expected in nature, we believe that the percentage recovery for ATP is more representative of the recovery of natural organic phosphorus by the test. In any event, the difference is small.

For nitrate, the efficiency of recovery is about 90% (after correction for column efficiency). This percentage slightly underestimates actual recovery of nitrates in unspiked samples, since addition of the spike increases the load on the reduction column, thus producing a small loss in recovery that is not characteristic of unspiked samples.

EPA Standards

Analysis of EPA standards showed good agreement of observed and expected values (Table 7).

4. Physical Variables and Major Ion Chemistry of the Lake

Geology

The geology of the Lake Dillon region has been well documented in connection with the construction of the Roberts Tunnel and the exploitation of mineral resources in the watershed. The publication of Wahlstrom and Hornback (1962) gives details and contains references to earlier studies.

Lake Dillon lies on a bed of stream and glacial deposits of Quaternary age, especially in the old riverbed, and on sedimentary rocks of Mesozoic age. Except near the valley bottoms, the Snake River drainage and small adjoining portions of the Blue River drainage are composed of igneous and metamorphic rock of Precambrian age. Nearer to the riverbed at lower elevations the Snake River drainage contains stream and glacial deposits of Quaternary age. In the upper Snake drainage there is an exposure of Tertiary quartzes (monzonite and monzonite porphyry).

There is a sharp break in the geology along a line that runs north and south to the east of the Blue River. This line represents the Williams Range Thrust Fault. To the west of this fault, we find Mesozoic sedimentary rocks and roughly equal exposures of Tertiary quartzes and felsite porphyry. Along the valley bottoms there are also extensive deposits of Quaternary age.

Morphometry

A bathymetric map of the lake is shown in Figure 5. It is useful to compare the central portion of the lake to the lake arms because differences between the arms and the

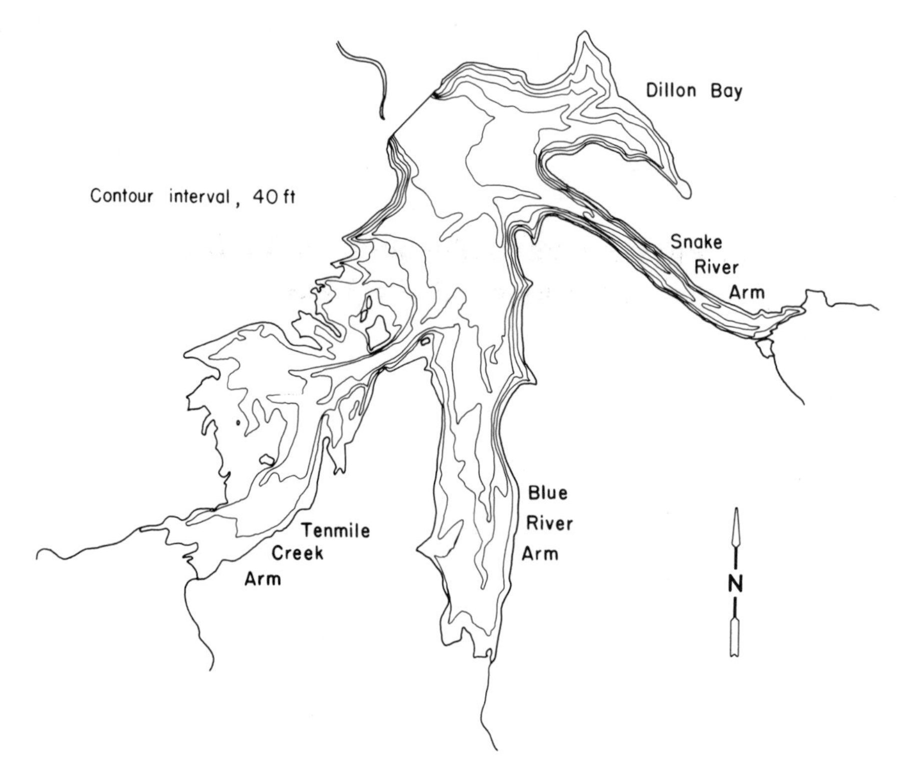

Figure 5. Bathymetric map of Lake Dillon.

main body of the lake will be analyzed in Chapter 9. Dillon Bay is relatively deep and opens broadly onto the main body of the lake. The Snake River arm is narrowest but has a greater mean depth than the other arms. The Blue River arm is shallow in its upstream half, where sediments are exposed at times of the year when drawdown is greatest. The Blue River arm opens broadly onto the main body of the lake, and this promotes exchange with the center of the lake. The Tenmile Creek arm has the greatest extent of shallow water; considerable areas of sediment surface are exposed in the arm during periods of extensive drawdown. Exchange between the main body of the lake and the Tenmile Creek arm could be restricted by the presence of islands near the mouth of the arm. Clearly the greatest opportunity for biological or chemical divergence from the main body of the lake is in this arm.

Figure 6 shows the depth-volume relationship; Figure 7 shows the hypsographic curve; Table 8 summarizes morphometric statistics related to water level. Although a small amount of storage above spillway level is possible, spillway level is for practical purposes the maximum lake level, so other depths are expressed in relation to spillway level. The Roberts Tunnel diverts water to Denver from the lake through an intake pipe 52 m below the spillway. The outlet to the Blue River is slightly below this, at a depth of 57 m when the water is at spillway level. The original maximum depth of the lake was 68 m, which would allow for just over 1% dead storage. The current maximum depth is probably somewhat less than this. The greatest maximum

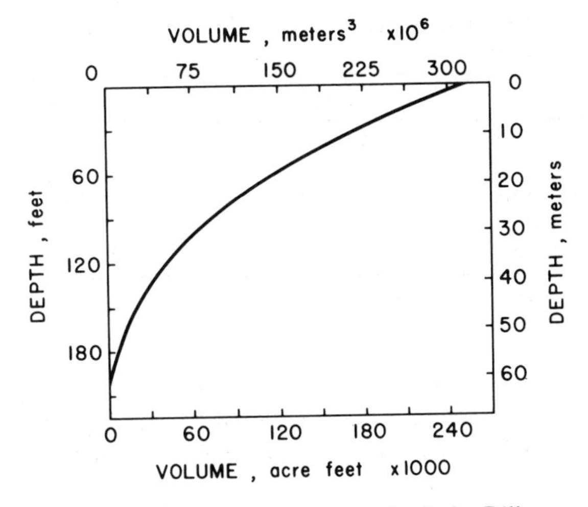

Figure 6. Depth–volume curve for Lake Dillon.

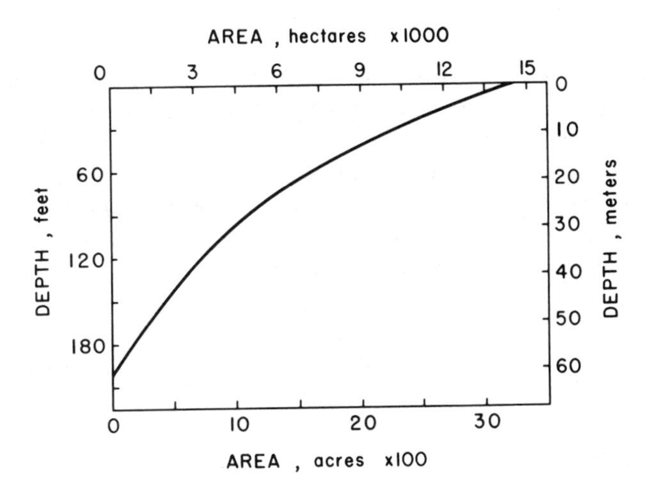

Figure 7. Hypsographic curve for Lake Dillon.

Table 8. Morphometric Statistics for Lake Dillon[a]

	Sea Level Reference		Spillway Reference	
	Feet	Meters	Feet	Meters
Spillway	9017	2748	0	0
Maximum storage (flood)	9025	2751	+8	+2.4
Roberts Tunnel	8846	2696	−171	−52
Blue River outlet	8829	2691	−188	−57
Original maximum depth	8795	2681	−222	−68

[a] Source: engineering documents provided by the Denver Water Department.

Table 9. Morphometric Statistics on Lake Dillon, Assuming Water at Spillway Level[a]

	English		Metric	
Lake volume	262,000	acre-feet	0.323	km^3
Lake area	3300	acres	13.35	km^2
Length of shore	24.5	miles	39.4	km
Mean depth	79.3	feet	24.1	m
Watershed area	212,900	acres	85,160	ha

[a] Source: Denver Water Department.

depth measured during the course of the study was 61 m but, because of the presence of the old river channel, some deeper spots are probably still present.

Table 9 summarizes the lake volume, area, length of shoreline, and mean depth when the water is at spillway level.

Hydrology

We summarize here only the changes in lake level and total flow into and out of the lake. A more detailed consideration of the hydrology of individual watersheds will be taken up later in connection with the analysis of nutrient income to the lake.

Figure 8 shows the total calculated inflow to the lake for the study period as estimated by the Denver Water Department. There are differences in the actual inflow and the calculated inflow because the calculated inflow is estimated as the sum of the outflow (which is measured) and the change in water volume (which is estimated from the depth-volume relationship). Errors in equating calculated inflow with actual inflow are not of concern here but will be discussed in Chapter 12. It will also be shown in Chapter 12 that groundwater inflow is at most 5-10% of the total inflow.

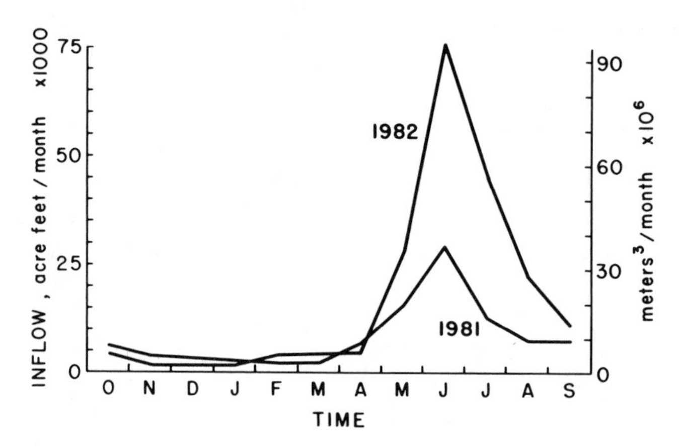

Figure 8. Calculated inflow as obtained by the Denver Water Department for Lake Dillon over the study period on a monthly basis (water years).

The calculated inflow to the lake for 1982 was about twice the amount for 1981. In both years the peak inflow occurred in the month of June, but the distribution of inflow around the peak was not quite the same in the 2 years. In 1981 the total inflow was quite symmetrical around the month of June, but in 1982 high flow extended well into July. In both water years the months November through March showed a stable minimum inflow of about 2000 to 4000 acre-ft per month.

Figure 9 shows the total calculated inflow since 1963 (calendar years). The purpose of this figure is to put the 2 years of the study into perspective hydrologically. For the period of record, 1981 was well below the average (only 1963 was lower) and 1982 was somewhat above the average (13 of the 19 other years are below 1982).

Figure 10 shows the total outflow per month for 1981 and 1982. Water leaving the lake follows some combination of three pathways: (1) the Roberts Tunnel, (2) the Blue River outlet, and (3) the spillway. Since the Roberts Tunnel and Blue River sub-

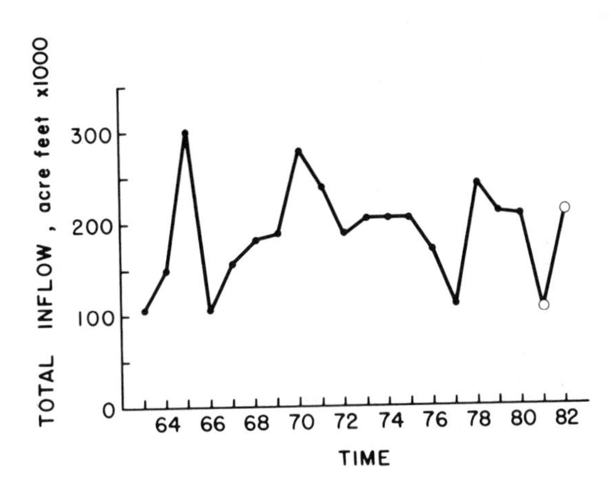

Figure 9. Calculated inflow since 1963 for Lake Dillon (Denver Water Department).

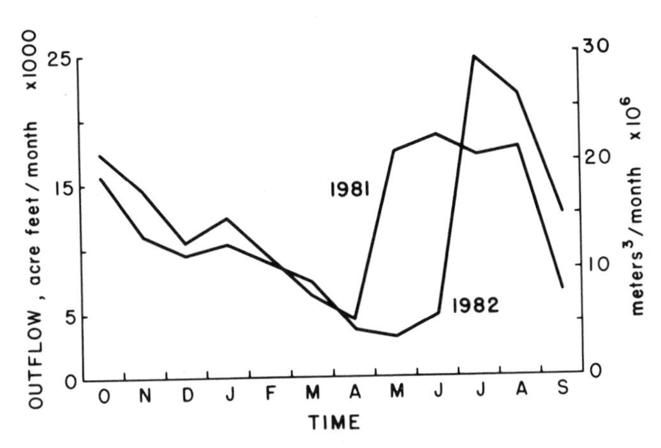

Figure 10. Outflow from Dillon over the study period on a monthly basis (Denver Water Department).

surface outlet remove water from points near the bottom of the lake, they have not been separated in Figure 10, nor need they be treated separately in any of the analyses. Spillway losses must be separated from other losses, however. In 1981 no water left the lake over the spillway, but in 1982 water left the lake over the spillway between 9 July and 4 November. Since water going over the spillway removes water from the epilimnion of the lake, the timing and extent of spillway flow are important. The Denver Water Department does not keep records of the apportionment of outflowing water between the spillway and the outlet pipe, although it is known that spillway flow typically dominates when the two occur together. Because the surface and deepwater nitrate and conductance differ considerably at the time of spillway flow, we can estimate the relative contribution of spillway outflow to total outflow as follows: $C_c = XC_t + (1 - X)C_b$, where C_c is the composite concentration of NO_3-N or conductance in the Blue River below the dam, C_t is the concentration at the top and C_b at the bottom of the water column, and X is the proportion of water going over the spillway. The values vary some from week to week, but the estimate shows that about 80% of the water reaching the Blue River between 9 July and 4 November 1982 came over the spillway. Using this percentage as a constant, we have computed the spillway loss on a weekly basis over the period of spillway flow (Table 10).

Figure 11 shows the change in lake level during 1981 and 1982. The lake is typically drawn down in the winter months to meet demand and legal requirements for minimum flow in the Blue River during the period of low winter streamflow. The rate of drawdown in the 2 years was essentially the same, but the effect on lake level was

Table 10. Estimated Partitioning of Outflow Over the Period of Spillway Loss (1982)

Week	Acre Feet Per Week			
	Total Outflow	Outflow from Bottom[a]	Outflow over Spillway	Spillway Outflow as %/wk of 0- to 5-m Layer
9 Jul–15 Jul	4478	3376	1102	2.5
16 Jul–22 Jul	2458	3951	2507	5.6
23 Jul–29 Jul	5024	3795	1229	2.8
30 Jul–5 Aug	7228	5546	1682	3.8
6 Aug–12 Aug	3162	679	2483	5.6
13 Aug–19 Aug	5396	1076	4320	9.7
20 Aug–26 Aug	5322	1064	4258	9.5
27 Aug–2 Sept	8942	1788	7154	16.0
3 Sept–9 Sept	3008	602	2406	5.4
10 Sept–16 Sept	3194	639	2555	5.7
17 Sept–23 Sept	2830	566	2264	5.1
24 Sept–30 Sept	2476	495	1981	4.4
1 Oct–7 Oct	2069	415	1654	3.7
8 Oct–14 Oct	1858	372	1486	3.3
15 Oct–21 Oct	1844	369	1475	3.3
22 Oct–28 Oct	1774	355	1419	3.2
29 Oct–4 Nov	1744	538	1206	2.7

[a] Includes both Roberts Tunnel and Blue River outflow pipe.

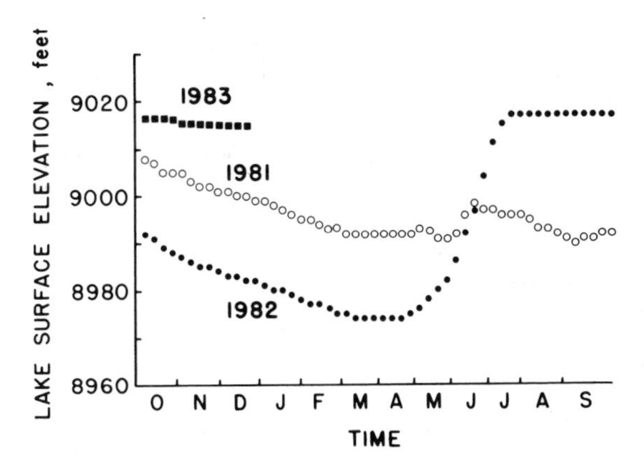

Figure 11. Change of lake level with time over the study period. Data points are mid-week values (Denver Water Department).

more drastic in the water year 1982 because the snowpack of water year 1981 was not enough to fill the lake to spillway level.

The combined effect of lake volume changes and changes in total inflow are illustrated in Figure 12 in terms of the percent of volume added to the lake per month for each month of the study. Dilution is most significant in June, less so in May and July, and minimal in other months.

Table 11 summarizes the flow of water into and out of the lake on an annual basis during 1981 and 1982. Both calendar year and water year figures are shown, but the difference between these two conventions is not very great.

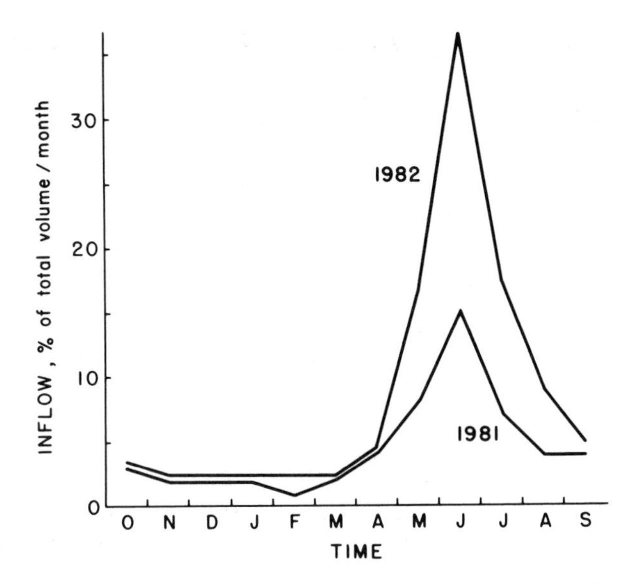

Figure 12. Volume of monthly inflow as a percentage of actual lake volume.

Table 11. Summary of Water Flow into and out of the Lake, Based on Daily Statistics Supplied by the Denver Water Department (Cumulated Monthly and Daily Statistics from the Same Source Differ Slightly)

	Calculated Inflow		Measured Outflow	
	Acre-ft	km^3	Acre-ft	km^3
1981				
Calendar Year	105,302	0.130	145,912	0.180
Water Year	106,193	0.131	152,565	0.188
1982				
Calendar Year	217,386	0.268	119,838	0.148
Water Year	210,255	0.259	131,989	0.163

Temperature

Lake Dillon is dimictic. The year can thus be divided into four intervals: (1) winter ice cover, (2) spring mixing (spring overturn), (3) summer stratification, and (4) fall mixing (fall overturn).

In both 1981 and 1982, ice cover formed on Lake Dillon in late December or early January (Figures 13 and 14). In both winters the maximum ice thickness reached about 0.5 m in midlake. In 1981 the ice was exceptionally transparent (i.e., without

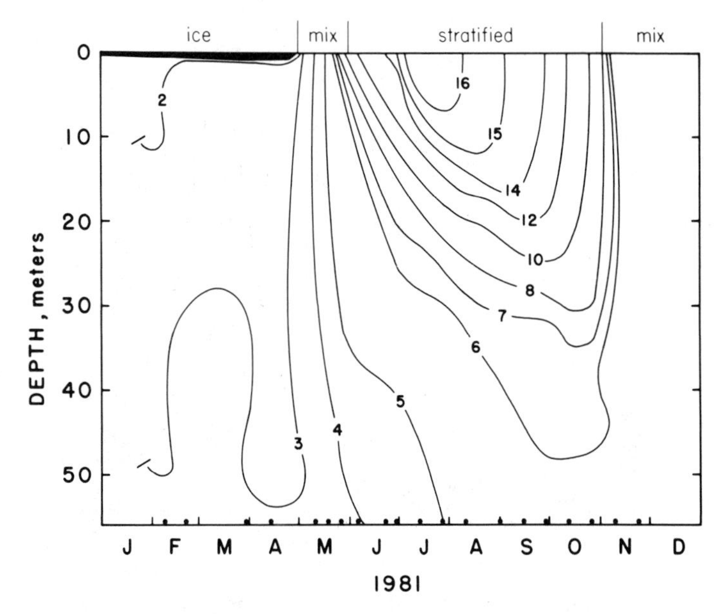

Figure 13. Time–depth diagram for temperature (°C), 1981.

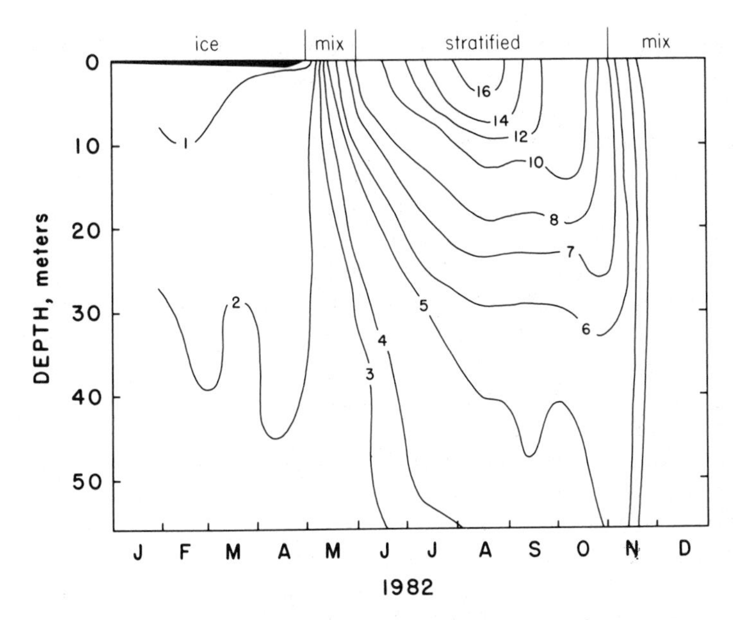

Figure 14. Time–depth diagram for temperature (°C), 1982.

bubbles or snow layers), and, because of the small amount of winter snowfall, was not covered with significant amounts of snow during the first 2 months of ice cover. Because an ice cover of this type transmits light almost as well as water (Ragotzkie 1978), unusually large amounts of light reached the water column during the winter of 1981. By April of 1981 light penetration was partially blocked by snow accumulation and thaw cycles that created an opaque granular layer at the top. In 1982 there was snow cover from the very beginning, and it lasted through the entire winter. Snow on the lake surface did not accumulate to great depth because of wind, but there were few completely open areas in the winter of 1982. Since snow has an absorption coefficient of about 0.24 cm^{-1} (Ragotzkie 1978), 10 cm of snow, which is near the average amount on the lake in 1982, would remove about 90% of the light.

In both 1981 and 1982, the ice cover broke up in early May. Temperatures under the ice in both years were between 1 and 3°C over most of the water column, and temperatures closer to the ice were slightly lower than near the bottom, as expected. Warming of the surface occurred slowly in both years after the ice cover disappeared, but especially so in 1981. As long as a water column is below 4°C, warming at the surface actually creates instability becuase of the increase in density of water as it is warmed from the freezing point to 4°C. Even after the water column warms to 5°C there is little resistance to mixing because of the very slight change in density of water in the vicinity of 4°C (Hutchinson 1957). Thus there is no tendency to stable layering until the surface water exceeds 5 to 6°C. Until then, heat is efficiently distributed through the entire water column by the wind, and the water thus warms only slowly at the surface.

In Lake Dillon some beginning indications of layering could be seen as the lake exceeded 5°C in the last half of May. Although stratification was very weak at first, complete mixing of the water column ended near the last of May in both years. June 1 thus marks the beginning of the stratification season.

By the last half of June in both years thermal layering had become very pronounced. The upper water column reached 12°C by the last half of June, and the thickness of the upper mixed layer at this time was about 5 m. We equate the upper mixed layer with the epilimnion, although the epilimnion is sometimes defined somewhat differently (Hutchinson 1957). The June thermocline was thick; it extended from 5 m down to about 20 m. In both 1981 and 1982 the upper water column continued to take up heat well after stratification was first established. In 1981 the lake reached maximum heat content in the last week of July, when the temperature of the epilimnion rose to just over 16°C. In 1982 the warming occurred slightly more slowly, producing maximum epilimnetic temperature in the first week of August, but the maximum epilimnetic temperature was essentially the same as that in 1981. In both years the thickness of the mixed layer remained 5-10 m for the first half of the stratification season, i.e., until the middle of August, but the average thickness was somewhat greater in 1981 than in 1982 (Figure 15). In both years cooling of the epilimnion began in the last half of August. Cooling of the upper water column reduced the stability of stratification sufficiently to allow gradual erosion of the thermocline by late August or early September. In 1981 the thickness of the mixed layer had reached 15 m by the middle of September and 25 m by the middle of October. In 1982, the mixed layer had thickened to 10 m by the middle of September and 25 m by the middle of October.

The stratification season ended with complete mixing during the first half of November in both 1981 and 1982. By the time complete mixing occurred, the lake had already cooled to a temperature of about 6°C. Isothermal cooling accompanied by complete mixing of the water column then proceeded through November and most of December until the establishment of ice cover at the end of December or beginning of January.

Table 12 gives a synopsis of the layering and mixing events in Lake Dillon in 1981 and 1982. There were no significant differences in the timing of major events in the 2 years. Elevation is obviously responsible for the extended ice cover, short stratifica-

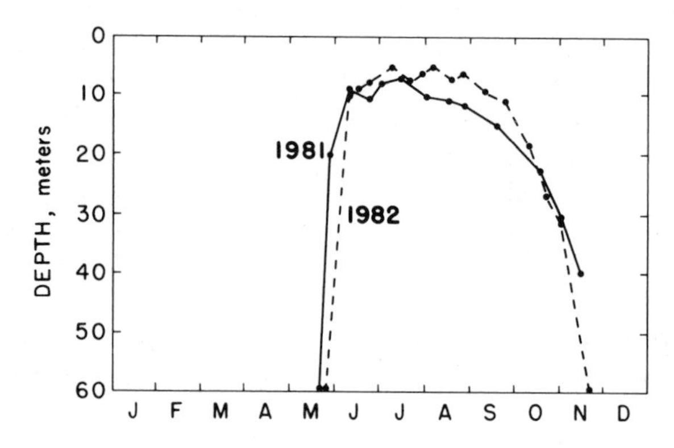

Figure 15. Thickness of the mixed layer during the ice-free seasons of 1981 and 1982.

Table 12. Seasons of Layering and Mixing in Lake Dillon

Season	Duration
Ice cover	January, February, March, April
Spring mixing	May
Summer stratification	June to first week of November
Fixed epilimnetic thickness	June, July, August
Increasing epilimnetic thickness	September, October
Fall mixing	Last 3 weeks of November, December

tion, and cool surface temperatures. According to the conversions suggested by Lewis (1983a), Lake Dillon's thermal and mixing behavior would be very similar to that of a sealevel lake at 14°C higher latitude (i.e., ~54°).

Transparency

Figure 16 shows the transparency of the lake over the 2-year study period both in terms of the secchi depth and the depth of 1% light. The depth of 1% light was approximated from the secchi depth as follows:

$$\eta_t = -\ln(0.19)/z_s$$

$$z_{0.01} = -\ln(0.01)/\eta_t$$

where η_t is the extinction coefficient of photosynthetically active radiation (350-700 nm, m^{-1}), z_s is secchi depth (m), $z_{0.01}$ is the depth at which 1% of the surface photosynthetically active radiation (PAR) is found (m). The approximation assumes that the secchi depth always falls at 19% of the surface irradiance. The basis of this assumption is a series of comparisons between light meter readings and secchi depth readings as reported in the methods section. The purpose of the conversion from secchi depth readings to depth of 1% light is to show the approximate depth at which photosynthesis could occur rapidly enough to match the respiration of algal cells (Talling 1971).

Figure 16 shows that significant photosynthesis in Lake Dillon was always limited to the top 15 m and was usually limited to the top 10 m of the water column. The main features of the seasonal change in transparency could be summarized as follows: (1) transparency was moderate to high under ice, (2) transparency was consistently low over the entire first half of summer stratification, and (3) transparency was consistently high during spring and fall mixing and over the last half of summer stratification. The readings under ice may be underestimates, especially in 1982 where there was significant snow cover. Because the readings were taken through a hole in the ice, light could not reach the secchi disk as well as it did under ice-free conditions.

Four factors can affect the transparency of the lake: (1) the extinction of light by pure water, (2) the extinction of light by organic materials dissolved in the water, (3) the extinction of light by chlorophyll, and (4) the extinction of light by organic

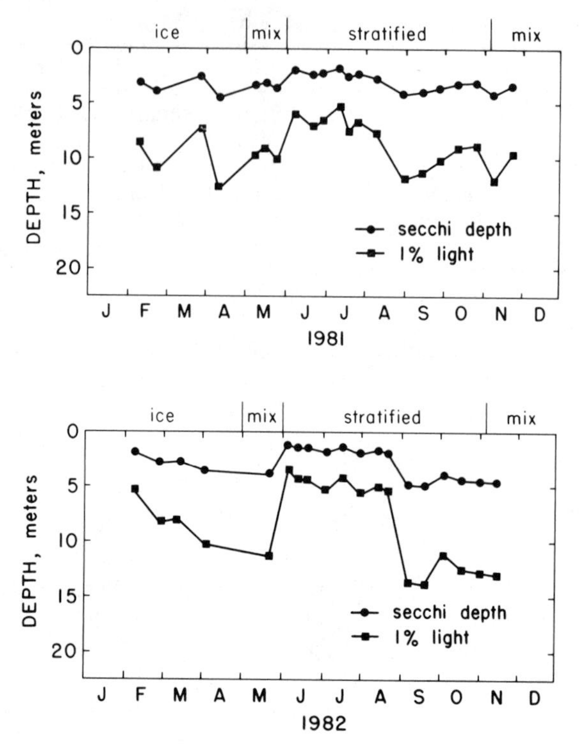

Figure 16. Secchi depth and depth of 1% light in Lake Dillon.

and inorganic particulate materials. The extinction coefficient (η_t) can thus be broken into components (Hutchinson 1957):

$$\eta_t = \eta_w + \eta_d + \eta_c + \eta_p$$

where w represents pure water, d represents dissolved substances, c represents chlorophyll, and p represents particles. η_t is the exponential coefficient in the equation that approximates the vertical attenuation of light:

$$I_z = I_o e^{-\eta_t z}$$

where z is depth (m), I is irradiance, and η_t has units m^{-1}. Values of η_t and its components are always somewhat approximate because of the change in spectral composition of light with depth (Smith 1968).

The effect of pure water is relatively constant through the seasons, although it does vary some because of seasonal changes in spectral composition of downwelling irradiance. A reasonable approximation of η_w in the euphotic zone of Lake Dillon would be 0.1 m^{-1}. Dissolved organic materials contribute to the extinction of light in the lake but are seasonally quite constant, as indicated by absorbances of filtered samples from Dillon. Dissolved organics, which are responsible for the extinction component η_d, are not present in large amounts in Dillon. From the absorbances at 360 nm and the equation of Lewis and Canfield (1977), we estimate the dissolved organic carbon as 1.5–1.6 mg/liter. From the absorbances of filtered water and the approximation method developed by Talling (1971), we estimate that, in complete absence of all particulates, including phytoplankton, the secchi depth would be at least 10 m. This is based on $\eta_w + \eta_d = 0.17\ m^{-1}$ which is a maximal figure, since η_w is smaller for more transparent lakes. Secchi depth of 10 m corresponds to a 1% light level of about 27 m. For comparison, the average secchi depth in Lake Tahoe, a lake exceptionally free of particulates and soluble organics, is about 28 m (Goldman 1974).

Separation of the effect of chlorophyll from the other effects on transparency involves some assumptions, but is feasible as an approximation. First it is useful to make a few qualitative observations concerning the relative importance of algae and nonliving suspended matter. In 1982 a major and precipitous decline in transparency occurred just at the onset of stratification but before chlorophyll levels in the upper water column had begun to climb (Figure 16). For example, the secchi depth of the lake in 1982 changed from 4.1 m on 24 May to 1.3 m on 7 June. The effect was observable at the same time in 1981 but was less dramatic. This quick decrease in transparency coincided with the spring runoff, and, since the chlorophyll had not increased much at this point, the sudden decrease in transparency must be attributable to nonliving particulate materials added to the lake by spring runoff. The transparency of the lake prior to the onset of runoff shows that mixing of the lake was not responsible for the decrease in transparency. All evidence thus points specifically to the major role of spring runoff in decreasing the transparency of the lake in early June.

Transparency remained low in the lake both years into August, despite the much earlier decline in amount of surface water entering the lake. Low transparency after June must therefore have been in large part attributable to the buildup of phytoplankton biomass. This is confirmed by data on chlorophyll, to be presented in the next chapter, and by microscopic examinations of samples, which showed a major decline in nonliving particulates after runoff had passed.

In the last half of the stratification season of both years, beginning in August, there was a return to high transparency. This was accompanied by drastic declines in the surface chlorophyll. The reasons for this return to higher transparency are related to the nutrition of the algal populations, so the complete explanation is deferred until the chlorophyll data are discussed.

Separation of the effect of chlorophyll from other effects on transparency can be achieved by estimation of the chlorophyll-specific absorption coefficient, e_s, which is related to the extinction coefficient for chlorophyll as follows: $e_s = \eta_c/B$, where η_c is extinction due to chlorophyll (ln units per meter), e_s is the chlorophyll-specific coefficient (ln units $\cdot$ m^2/mg Chl*a*) and B is chlorophyll *a* concentration (mg Chl*a*/m^3). e_s is traditionally treated as a constant for a given lake, although it is known to vary some with depth (due to changes in spectral composition) and with seasonal changes in phytoplankton composition (Kirk 1975). The estimate of e_s for Lake Dillon is based on linear regression of extinction coefficient (dependent variable) on chlorophyll concentration (independent variable), which yields a slope equal to e_s. The slope of the regression for Lake Dillon is 0.016 m^2/mg. This is within the midrange of literature values, which generally fall between 0.008 and 0.021 m^2/mg and show a mode around 0.015 (Talling 1982).

From the value of e_s and the chlorophyll concentration on any given day, the amount of light extinction due to chlorophyll (η_c) can be approximated. If the effect of pure water and dissolved organics, assumed constant at $\eta_w + \eta_d = 0.17\ m^{-1}$, is added to η_c, the remaining extinction must be the nonchlorophyll particulate effect, the last of the four components of extinction. The separation made in this way is shown in Figure 17.

Figure 17 shows that the contribution of water and dissolved substances to extinction was greatest at times of greatest transparency, as might be expected. Particulates were always an important component of extinction. This was especially so in early stratification, when microscopic examination showed abundant inorganic particulate material from runoff. This inorganic fraction sedimented almost completely by mid-July, yet the particulate contribution to extinction remained high, presumably because of biogenic particulates. Chlorophyll made an important contribution in winter, in midstratification, and in the fall after the mixed layer began to thicken.

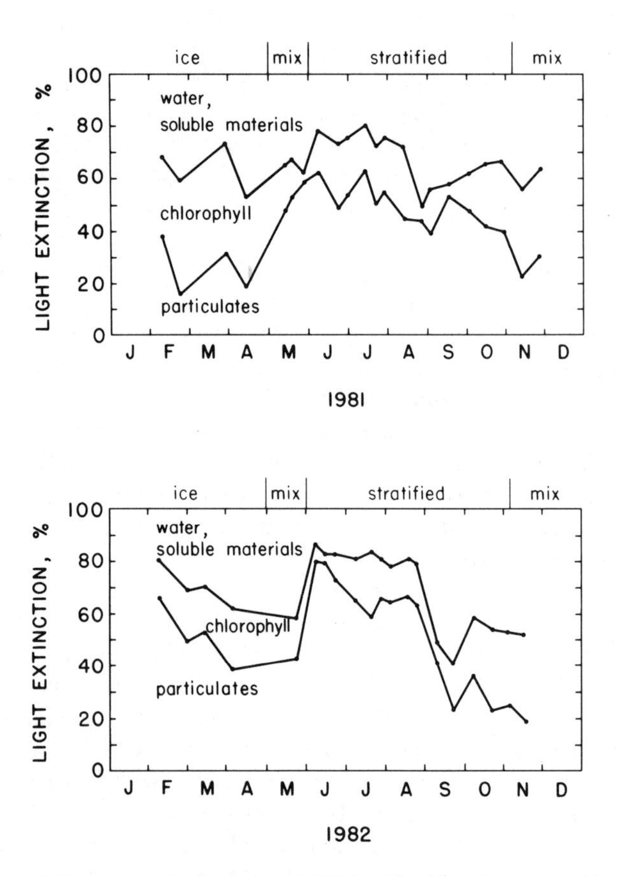

Figure 17. Cumulative percent of observed light extinction due to particulates, chlorophyll, and water plus soluble material fractions in Dillon for 1981 (above) and 1982 (below).

Historical Changes in Transparency

There is some historical information on the transparency of Lake Dillon. The secchi depth was measured by the National Eutrophication Survey in 1975. In August 1975, the mean of secchi depth measurements at several sites on the lake was 7.8 m. In September of the same year the mean was 8.3 m. Since the secchi depths at the same times in 1981 and 1982 were only about half these values, it would appear that a major decrease in transparency occurred between 1975 and 1981. Secchi depth data are also available from the Colorado Division of Wildlife (W. Nelson, personal communication). Nelson's values average lower than readings obtained at the same time by the Eutrophication Survey or the present study, possibly because of differences in discs or in the location of data collection. However, his series spanning 1975-1982 for secchi depth on or near August 25 shows some evidence of a trend toward lower transparency: 1975, 4.6 m; 1978, 3.6 m; 1979, 2.0 m; 1980, 2.3 m; 1981, 3.1 m; 1982, 1.6 m.

Composition of Major Ions

Table 13 shows the composition of major ions in water from Lake Dillon. There are some seasonal changes in major ions, particularly associated with the runoff in the month of June, but these are not great enough to have much biological significance, so the amounts of major ions were not quantified on a routine basis. Table 13 indicates that the water of Lake Dillon is a bicarbonate type exceptionally rich in sulfate. Cations are dominated by calcium. The total ionic strength, as indicated by Table 13

Table 13. Major Ion Composition of Lake Dillon, Based on Analysis of a Sample Taken at the Index Station, 4 October 1982

	Concentration	
	mg/l	meq/l
Cations		
Ca^{++}	21.32	1.068
Mg^{++}	3.58	0.295
Na^{+}	3.45	0.150
K^{+}	0.54	0.014
Total cations	28.89	1.528
Anions		
HCO_3^-	47.10	0.772
$SO_4^=$	39.10	0.814
Cl^-	3.90	0.110
NO_3^-	0.93	0.015
Total anions	91.03	1.711
Total ions	119.92	3.239

and by the conductance measurements that were made routinely throughout the course of the study, is somewhat higher than might be expected for an undisturbed Rocky Mountain watershed of similar geology. The total ionic strength and ion composition of the lakewater have been augmented to some extent by mining and dispersed earth disturbance, as will become evident in the analysis of stream chemistry. Expected conductance computed from Table 13 and ionic mobilities is 179 μmho/cm at 25°C, which checks well with the observed conductance (170 μmho/cm).

Conductance

Conductance of lakewater was slightly higher in 1981 than 1982, as shown in Table 14, which gives annual averages for the top, middle, and lower water column. The difference between years was more pronounced in the upper water column than in the lower water column. Conductance showed significant seasonal variation and a degree of vertical layering under ice and during summer stratification. These features are illustrated by Figure 18, which shows the change of conductance through time at the top of the water column (0-5 m) and in the bottom 5 m for the 2 years of the study. In both years under ice the surface water was lower in ionic solids than deep water. This layering must come into being after ice cover, since fall mixing homogenizes conductance. Retention of melt water near the surface is the probable explanation. Surface temperatures under ice were in the vicinity of 0°C at the top and, since water is less dense at 0°C than at slightly higher temperatures, there would have been a slight resistance to mixing to help retain the layering as long as the ice was protecting the lake surface from wind.

Vertical homogeneity of conductance was typical of the spring mixing period, as expected. With the onset of thermal stratification in June, there once again was a divergence in conductance of surface water and deep water. The divergence appeared rather suddenly, suggesting that spring snowmelt was at first held in greatest amounts

Table 14. Means for Conductance, pH, and Alkalinity at Three Depths in the Water Column at the Index Station

	Depth (m)		
	0-5	20-25	35-40
Conductance (μmho/cm)			
1981	168	167	178
1982	152	153	172
pH			
1981	7.45	7.36	7.24
1982	7.80	7.59	7.54
Alkalinity (mg/l CO_2)			
1981	33.9	34.4	34.3
1982	34.5	34.3	35.4

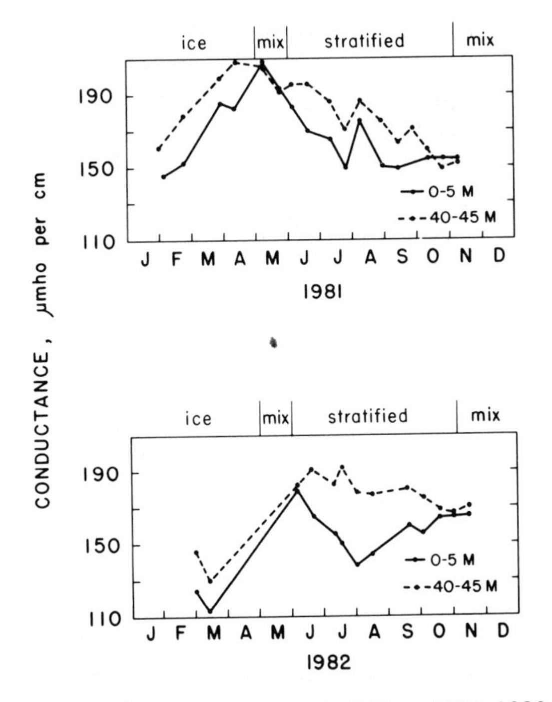

Figure 18. Conductance in Lake Dillon, 1981–1982.

in the upper water column after thermal stratification began. Not all of the runoff entered the upper water column, however, because the lower water column also showed steady dilution. Furthermore, the conductance of the upper water column was homogeneous down to approximately 30 m, a depth much greater than the upper boundary of the thermocline. Thus the dilution caused by spring snowmelt affected the entire water column but was more pronounced in the upper 30 m than in the lower 30 m. This accounts for the higher annual average conductance of the lower water column (Table 14). With the fall mixing, vertical homogeneity of conductance was established as shown in Figure 18.

It is possible to estimate the depth at which entering streams penetrated the lake, since a stream will tend to seek the layer of its own density. This is not an important issue when the lake is mixing freely, but is potentially of interest during the stratification season when the entering water may tend to flow selectively into the upper, middle, or lower water column. In order to make such an estimate, we computed the density profile of the water column on each sampling date during stratification. Taking into account not only temperature, but also dissolved and suspended solids, we also calculated the density of the three rivers on each date. By comparison we could then show the depth at which river and lake density would be matched, and this is the depth to which the rivers would tend to flow. The results are shown in Figure 19. In both years the bulk of flow was predominantly into the upper middle of the water column, and in both years the depth of penetration increased from June to October.

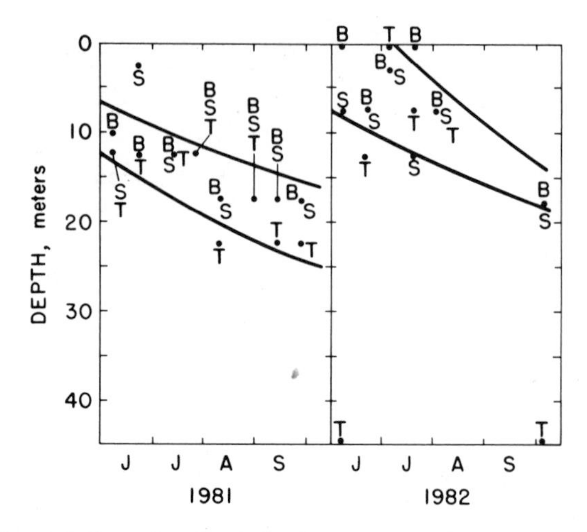

Figure 19. Calculated inflow depth of the Snake River (S), Blue River (B), and Tenmile Creek (T) during stratification in 1981 and 1982.

In 1981 the penetration began at about 10 m, just below the thermocline, and increased to about 25 m by late stratification (due mainly to warming of the lakewater). In 1982, penetration was not so deep; it began at about 5 m, and slowly passed down to about 15 m. The larger flow of cold water to the metalimnion and continual heat

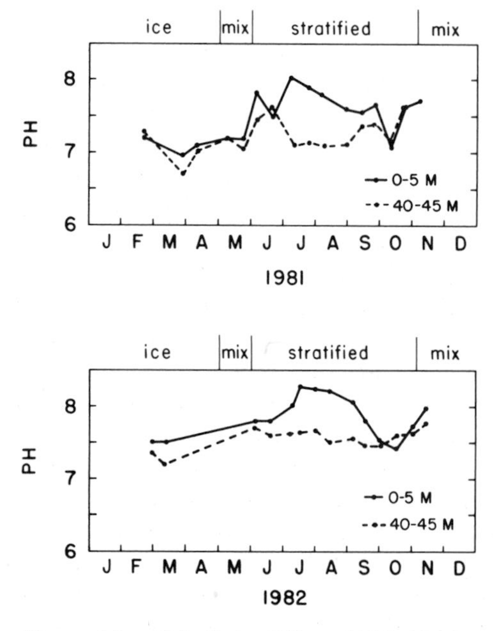

Figure 20. pH in Lake Dillon, 1981–1982.

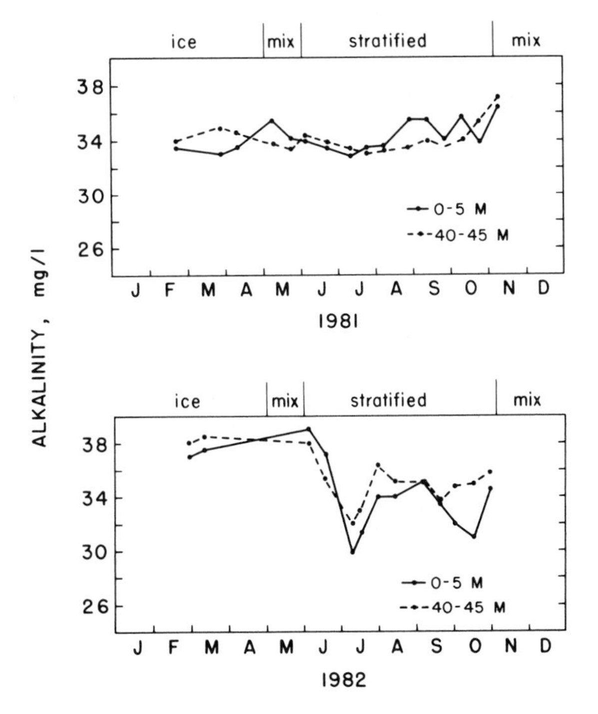

Figure 21. Alkalinity in Dillon, expressed as milligrams per liter of CO_2.

loss over the spillway in 1982 probably explain why the mixed layer was thinner in 1982 than in 1981.

pH and Alkalinity

Table 14 shows that the pH at all depths was slightly higher in 1982 than in 1981, and that in both years the mean pH at the bottom of the water column was slightly lower than at the top, presumably due to the influence of photosynthesis near the top of the water column. Figure 20 shows the seasonal patterns of pH near the surface (0–5 m) and in deep water (40–45 m). The pH levels at top and bottom diverged in a major way only after rapid photosynthesis began in late June. Photosynthesis elevates pH in proportion to the rate of removal of free CO_2; this explains why surface pH has a seasonal pattern similar to that of chlorophyll. The pH levels were at no time extraordinary, however, and do not suggest extreme depletion of free CO_2, which is significant only above pH 9.0 (Talling 1976). In both years there was a return to lower pH levels as photosynthesis dropped off after July, and a resurgence after nutrient depletion was relieved by thickening of the mixed layer in fall.

Alkalinity values fell within a relatively narrow range, as shown by Table 14. Seasonal effects were slight (Figure 21), but the large runoff of June 1982 can be identified as the cause of a reduction in alkalinity.

5. Phosphorus and Nitrogen in Lake Water and Sediments

Phosphorus and nitrogen are the two elements most likely to limit phytoplankton growth. Detailed information on the concentrations, vertical distributions, and chemical fractions of these elements is therefore useful in interpreting average phytoplankton abundances and seasonal changes in abundance. We present here an overview of the phosphorus and nitrogen chemistry that can be referenced in later chapters dealing with phytoplankton.

Phosphorus in Lake Water

The concentration of total phosphorus and the contributions of soluble reactive phosphorus (SRP), soluble organic phosphorus, and particulate phosphorus to this total are summarized in Table 15. The average SRP values were very low in both years and showed virtually no trend with depth. The averages for 1982 were slightly higher, however. This is explained mainly by the greater flow of water into the lake in 1982, as described more fully below in connection with seasonal trends. Soluble organic P was similar to SRP in showing no real trend with depth in the averages and in the slightly higher average for 1982. Particulate P, in contrast, was essentially the same the 2 years and showed a slight tendency toward higher values near the surface both years, probably due to the ability of phytoplankton, which are located mostly near the surface, to sequester phosphorus. As a result mainly of higher soluble P levels, total P concentrations were slightly higher in 1982 than in 1981. Slightly less than half of the total P was soluble on the average.

Table 15. Mean Concentrations of Total Phosphorus and Phosphorus Fractions at Various Depths between the Surface and Bottom at the Index Station[a]

	Concentration (μg/l)							
	Soluble Reactive P		Soluble Organic P		Particulate P		Total P	
Stratum	1981	1982	1981	1982	1981	1982	1981	1982
0–5	0.75	1.50	2.08	2.36	4.80	5.54	7.63	9.15
5–10	0.56	0.94	1.55	2.49	4.98	4.96	6.74	8.20
10–15	0.77	1.01	1.77	4.96	4.46	4.87	7.01	10.64
15–20	0.62	1.30	2.40	2.57	4.31	4.37	7.33	8.01
20–25	0.47	1.91	1.56	1.84	4.15	3.99	6.19	7.45
25–30	0.63	1.65	2.00	1.53	3.76	3.54	6.39	6.16
30–35	0.57	1.66	3.52	3.21	3.68	3.58	6.31	8.19
35–40	0.64	1.30	2.04	2.11	3.51	3.38	5.83	6.56
40–45	0.72	1.51	1.68	1.31	3.70	3.83	6.23	6.40
45–50	0.72	2.10	1.81	1.17	3.47	3.27	6.68	6.80
50–55	0.79	2.16	2.60	1.48	4.00	2.73	6.03	6.83
Mean	0.66	1.55	2.09	2.28	4.07	4.00	6.58	7.67
Mean %	10	20	31	29	60	51	100	100

[a] Fractions may not equal totals exactly due to occasional missing values for individual fractions.

Table 16 shows the results of a two-way analysis of variance (ANOVA) in which the total P and P fractions were tested for statistically significant (i.e., nonrandom) differences between years and depths. The 0- to 5-m layer was taken as representative of surface water and the 40- to 45-m layer was taken as representative of deep water. These two layers were chosen for the comparison because the 0–5 m layer is the only one that falls wholly within the epilimnion in summer and the 40–45-m layer is as far within the hypolimnion as possible without being on the bottom at the time of maximum drawdown. The years differed significantly only for SRP and total P, and depths differed significantly only for particulate and total P. No interactions of year with depth were significant.

Seasonal changes in the total phosphorus concentration and in the contribution of the different phosphorus fractions in the upper water column (0–5 m) are summarized

Table 16. Results of a Two-Way ANOVA Testing for Differences in Means between Depths (0–5 versus 40–45 m) and between Years (1981 versus 1982)

	Soluble Reactive P	Soluble Organic P	Particulate P	Total P
Difference between depths	Not significant	Not significant	Highly significant[a]	Highly significant[a]
Difference between years	Highly significant[a]	Not significant	Not significant	Significant[b]

[a] $P<0.01$.
[b] $P<0.05$.

in Figure 22. Comparable information is given in Figure 23 for the lower water column (40–45 m). There are some notable trends both in total phosphorus and in contributions of the fractions, especially in the surface layer. These trends can be understood for the most part in terms of the seasonal events occurring in the lake.

Spring runoff coincided with some of the highest total phosphorus concentrations in both years. Rising levels of total phosphorus in spring were coincident with an increase in the contribution of particulate phosphorus. We attribute this effect to the transport of large amounts of particulate material from the watershed at the time of runoff. Silt, which is an effective adsorption substrate for phosphorus (Golterman 1975), is particularly likely to make a major contribution to the particulate fraction of phosphorus at this time of the year. The peak in total phosphorus associated with runoff was considerably larger in 1982 than in 1981 when the runoff was lower.

From June through August, the interval over which the mixed layer is of stable thickness at 5-10 m, the total P concentrations of the mixed layer showed a persistent and dramatic decline indicative of sedimentation. Although the processes regulating total P in the mixed layer are by no means simple, the decline in both years is surprisingly well approximated by a straight line. In 1981 the decline of total P was about 0.10 μg/liter/day and in 1982 it was slightly higher (0.12 μg/liter/day). In both years the decline was halted by thickening of the mixed layer in September.

Particulate phosphorus was not brought to the surface from the lake bottom in significant amounts at the time of fall overturn, despite complete mixing of the water

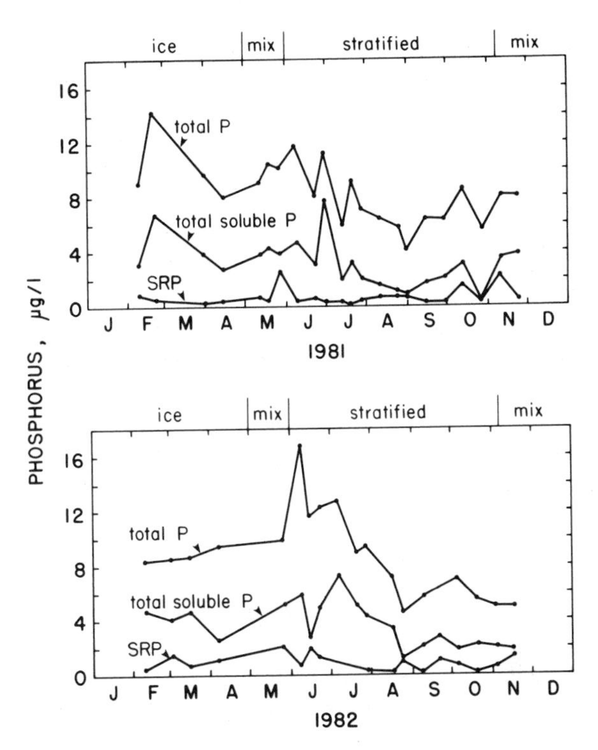

Figure 22. Total phosphorus and phosphorus fractions in the top 5 m of Lake Dillon.

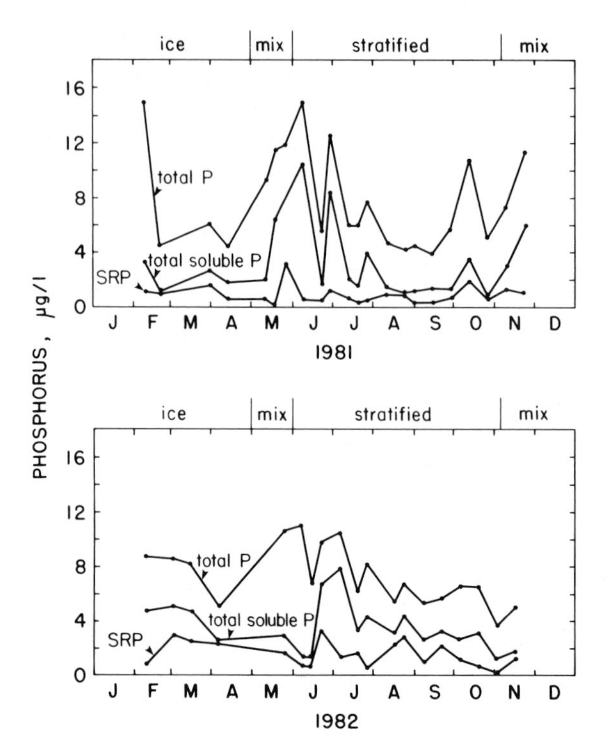

Figure 23. Total phosphorus and phosphorus fractions in deep water (40–45 m) of Lake Dillon.

column. From August until ice cover returned at the end of December, the total phosphorus hovered in the vicinity of 6 to 8 μg/liter, most of which was particulate and was undoubtedly tied up in phytoplankton and bacteria.

Soluble reactive phosphorus was seldom a major contributor to total phosphorus in either 1981 to 1982. SRP levels were typically in the vicinity of 1 μg/liter at the surface and were somewhat higher near the bottom during stratification (but not as an annual average: Table 15). Higher SRP at depth during stratification probably reflects lower biological demand there.

SRP concentrations occasionally rose significantly in the surface water. In May of 1981, SRP rose to about 4 μg/liter, a concentration well above the annual average. It seems almost certain that this rise in SRP was caused by spring mixing, when the winter phytoplankton had been dispersed from the surface and the spring phytoplankton had not been able to start surface accumulation because of the movement of the entire water column. Unfortunately there was only one May sample for 1982 because of a ring of ice around the lake that prevented launching of the boat during most of May. There were also some minor but consistent increases in the amount of surface SRP between September and December of both years. These probably were caused by the incorporation of hypolimnetic water with higher SRP into the mixed layer as the thermocline descended.

Table 17. Comparison of Total P Values from Various Studies of Dillon (Main Body of the Lake, Surface Samples)

Year	Total P (μg/l)			
	June	July	August	September
1975[a]	–	–	6	9
1977[b]	5	6	60[d]	–
1978[b]	7	6	6	–
1981[c]	11	7	6	6
1982[c]	12	10	6	6

[a] National Eutrophication Survey.
[b] Summit County/EPA study.
[c] Present study.
[d] Probably erroneous.

Soluble organic phosphorus made a significant contribution to total P at all times of the year except at the end of the stratification season. There was a definite and consistent decline in the total amount and percentage of soluble organic phosphorus at the surface between the first of July and the end of August both years. This was probably caused by strong biological demand for phosphorus, especially after the establishment of large phytoplankton populations by the end of June. Bacteria may have been involved as well, since the downward trend was also notable below the thermocline. Over a period of time, the phytoplankton and bacteria populations seemed to be able to sequester most of the soluble phosphorus, including the organic fraction, in the form of biomass. Alkaline phosphatase on cell surfaces may account for this ability to tap the soluble organic pool (Fogg 1975, Nalewajko and Lean 1980).

Table 17 summarizes the historical information on surface total P in the main body of the lake. The late summer values show no evidence of change, but the early summer values trend upward. The initial spring phosphorus, which is responsible for the growth burst of early summer, has increased, but the epilimnetic supply is ultimately depleted by sedimentation to much the same levels now as in 1975, according to the table.

Nitrogen in Lake Water

Table 18 summarizes the information on total nitrogen and nitrogen fractions at the index station. As would be expected in view of the presence of substantial amounts of dissolved oxygen at all depths throughout the year, the amounts of nitrite were exceedingly low, and nitrate was the dominant over ammonium among the soluble inorganic forms. There was a substantial depth gradient of average nitrate concentrations in both years. The seasonal analysis will show that this was caused by the removal of nitrate by phytoplankton in the surface waters during the summer. In 1981 and 1982 the nitrate concentrations were very similar in the upper half of the water column, but concentrations in the deepest water were consistently greater in 1982 than in 1981. Nitrate accounted for most of the soluble inorganic nitrogen pool both years and made up 30–45% of the total nitrogen.

Table 18. Average Total Nitrogen and Nitrogen Fractions at Different Depth Strata in 1981 and 1982 (Index Station)[a]

Depth Stratum (m)	Concentration (μg/l)													
	Nitrite N		Nitrate N		Ammonium N		Soluble Organic N		Total Soluble N		Particulate N		Total N	
	1981	1982	1981	1982	1981	1982	1981	1982	1981	1982	1981	1982	1981	1982
0–5	1.5	2.5	73	67	19	15	127	306	221	391	28	30	248	422
5–10	1.7	1.9	76	66	14	19	104	384	196	472	(28	30)[b]	224	502
10–15	1.2	2.3	74	67	13	15	104	216	192	300	(28	30)[b]	220	329
15–20	1.2	2.3	83	79	13	22	135	286	232	388	(28	30)[b]	260	418
20–25	1.2	3.8	101	106	18	33	109	165	230	308	(28	30)[b]	258	338
25–30	1.3	4.3	118	139	18	32	107	196	245	372	(28	30)[b]	273	402
30–35	1.2	5.0	155	174	15	30	108	243	280	452	(28	30)[b]	308	482
35–40	1.1	3.6	166	188	13	26	126	252	311	470	(28	30)[b]	339	500
40–45	1.1	2.9	187	224	13	26	113	227	315	480	(28	30)[b]	343	510
45–50	1.3	2.8	202	259	13	16	222	259	438	537	(28	30)[b]	466	565
Mean	1.3	3.1	123	137	15	23	126	253	266	417	(28	30)[b]	294	447
Mean %	.4	1	42	31	5	5	43	57	91	93	10	7	100	100

[a] Slight inconsistencies in the sums are due to occasional missing values for individual fractions.
[b] Values below 5 m set equal to annual mean for 0–5 m.

Ammonium nitrogen was present in measurable but very low amounts equal to only 5% of total nitrogen. Preferential uptake of ammonium by phytoplankton and continual oxidation of ammonium by microbes probably held the ammonium concentrations at a low level.

Soluble organic nitrogen was always an important component of the total nitrogen pool. In 1981 its percentage contribution to total nitrogen was slightly lower than in 1982; the grand average for the 2 years was about 50%. Absolute levels of soluble organic nitrogen in 1982 were considerably higher at all depths than in 1981. This is probably explained by the larger amount of water entering the lake in 1982. It will be evident from the analysis of river chemistry in Chapter 11 that river water entering Lake Dillon contains large amounts of soluble organic nitrogen, much of which is probably quite refractory, and that the concentrations increase with amount of runoff.

Table 18 shows that soluble nitrogen, including both organic and inorganic fractions, accounted for slightly more than 90% of the total nitrogen as an annual average. Particulate nitrogen was only measured in the surface water (0–5 m); the amounts at other depths have been set equal to the mean in the surface water. This is only an approximation, but errors associated with the approximation are not likely to be important in the computation of total nitrogen because of the small contribution of particulate nitrogen to the total. Total nitrogen, which is shown in the last two columns of Table 18, was higher in 1982 than in 1981, as would be expected from the patterns in the most important nitrogen fractions.

Table 19 summarizes the results of a two-way ANOVA similar to the one that was carried out on the phosphorus data. The ANOVA confirms statistically the difference between years in concentration of soluble organic nitrogen and total nitrogen. The difference between years in nitrate was not significant because it appeared only in deep water rather than over the entire water column. The ANOVA confirms statistically the differences in mean nitrate concentration with depth.

Figure 24 shows the seasonal changes in concentration of all nitrogen fractions at the surface of the lake. Some missing values for individual fractions have been filled in by linear extrapolation. Figures 25 and 26 show the complete details over time and depth of the distribution of nitrate, which is the most informative nitrogen fraction from the viewpoint of phytoplankton nutrition.

Figure 24 shows that ammonium was highest at the surface under ice. This probably reflects the lower demand for inorganic nitrogen and lower efficiency of

Table 19. Results of a Two-Way ANOVA Testing for Differences in Means between Depths (0–5 m versus 40–45 m) and between Years (1981 versus 1982)

	NO_3-N	NH_4-N	Soluble Organic N	Total N
Difference between depths	Highly significant[a]	Not significant	Not significant	Not significant
Difference between years	Not significant	Not significant	Highly[a] significant	Highly[a] significant

[a] $P<0.01$.

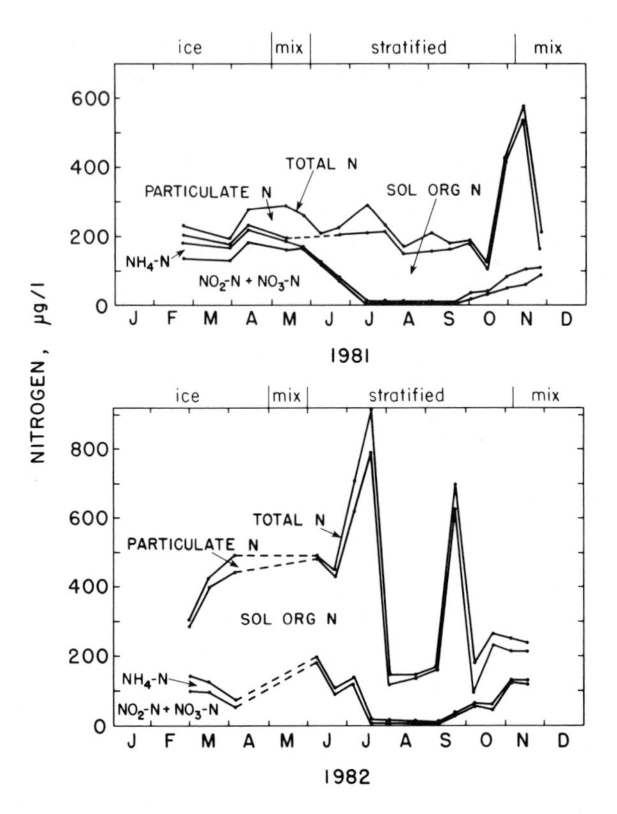

Figure 24. Total nitrogen and nitrogen fractions in the top 5 m of Lake Dillon.

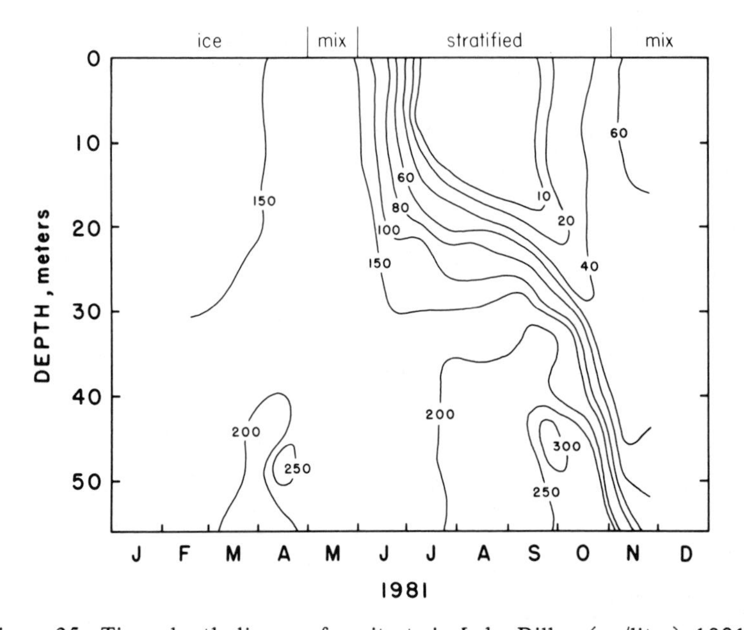

Figure 25. Time–depth diagram for nitrate in Lake Dillon (µg/liter), 1981.

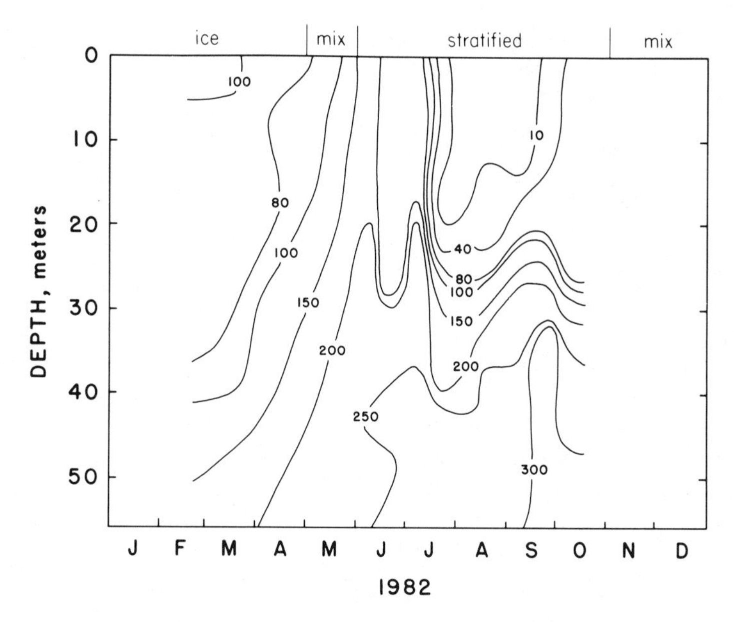

Figure 26. Time-depth diagram for nitrate in Lake Dillon (μg/liter), 1982.

nitrification at this time of year. Nitrate concentrations were highest in winter and spring. A spring increase was especially pronounced in 1982, probably because of the larger runoff that year. The inorganic nitrogen fractions began a precipitous decline at the lake surface as soon as stratification occurred. There is little doubt that this was caused by phytoplankton demand for inorganic nitrogen. Complete depletion of inorganic nitrogen at the surface occurred by the middle of July in 1981 and by the beginning of August in 1982. In both years, inorganic nitrogen began to climb again in September, in coincidence with thickening of the mixed layer. The seasonal trends are especially obvious from the time-depth diagrams for nitrate (Figures 25 and 26). Both figures show clearly that the nitrate depletion was associated specifically with the growth zone and did not occur below the thermocline where biological demand was much lower.

Soluble organic nitrogen peaked very sharply in 1982 late in the runoff period, suggesting that a major amount of nitrogen entering with the peak runoff was in soluble organic form. In 1981, a year of low runoff, the organic fraction was lost, so we do not know the nature of the runoff spike.

During summer stratification of both years there was a reduction in the organic nitrogen pool at the surface. The decline was so drastic in 1982 as to suggest the importance of abiotic processes, such as settling of flocculated organic matter. There was no evidence, however, of a steady drain on the soluble organic nitrogen fraction later as phytoplankton completely depleted the inorganic nitrogen. Thickening of the epilimnion in fall restored high levels of soluble organic nitrogen as soluble organic matter was brought to the surface from the hypolimnion.

Table 20 summarizes the total nitrogen and total phosphorus data as an average over the entire water column for both years and shows the molar ratio of nitrogen to

Table 20. Summary of Average N and P Amounts for the Whole Water Column at the Index Station

	1981	1982	Composite
Total N (μg/l)	294	447	370
Total P (μg/l)	6.6	7.1	6.8
Total N (μmole/l)	21	31	26
Total P (μmole/l)	0.21	0.23	0.22
N:P ratio (molar)	100	134	118

phosphorus. The ratio of nitrogen to phosphorus is exceptionally high. Municipal sewage typically has a ratio of 6 to 14, and most inland waters fall between 15 and 40 (Stumm and Baccini 1978). The very high ratios of Dillon are partly explained by tertiary treatment for P, which greatly raises the N to P ratio of effluents. The ratio of N to P for phytoplankton biomass is typically between 10 and 17. Thus the ratios of Dillon are suggestive of incipient phosphorus limitation. Differences in vertical distribution of N and P and in the differential biological availability of N and P fractions require other and more direct evidence of phosphorus control, however.

Sediment Phosphorus

Sediment samples from eight different sites were analyzed for particulate phosphorus and interstitial phosphorus. The collection sites for sediment included the heads and mouths of the lake arms as well as the main body of the lake. The wet surface sediments from a given site were mixed and a subsample equivalent to 10 to 20 mg of dry material was dried to constant weight in a tared container and then analyzed for total phosphorus by the particulate phosphorus procedure described in Chapter 3. Another subsample of wet sediment was centrifuged for 30 min at 8000 rpm, after which 5 ml of the supernatant was withdrawn and analyzed for total soluble phosphorus according to the procedures used for lake water. The results of these analyses are reported in Table 21.

There was surprisingly little variation in the phosphorus per unit dry weight of sediment. Such variation as can be discerned in the data cannot be related in any way to particular lake arms or to particular depths. Thus it appears that the lake bottom can be treated as having 0.1% phosphorus on the average with random variation spanning about ±0.02%. The percentage phosphorus in the dry sediments of the lake is not especially high. For example, the summary data of Brunskill et al. (1971) show an average of 0.12% for 23 Wisconsin lakes and 0.11% for 30 eastern Ontario lakes.

The total soluble phosphorus in interstitial water was high by comparison with the overlying lakewater, as expected from the decomposition of organic matter liberating phosphorus in the sediments. These values, although much higher than lakewater, are not extraordinary, even for relatively unproductive lakes. For example, the summary of Brunskill et al. (1971) on interstitial water chemistry for unproductive lakes of the Canadian Shield shows a range of two orders of magnitude in the interstitial total soluble phosphorus and a mode of 220 μg/liter P. In Lake Dillon the diffusion gradient

Table 21. Particulate P and Interstitial P from Surface Sediments Collected July 26, 1982 at Eight Sites on Lake Dillon

Collection Site	Water Depth (m)	Sediment P (% P in Dry Sediment)	Interstitial P (Total Soluble, μg/l)
Lake Center (index)	60	0.085	203
Mouth, Dillon Bay (main)	32	0.102	113
Mouth, Snake Arm (main)	38	0.092	103
Mouth, Blue Arm (main)	30	0.103	152
Mouth, Tenmile Arm (main)	33	0.115	132
Head, Snake Arm (survey 2)	23	0.093	136
Head, Blue Arm (survey 4)	18	0.124	132
Head, Tenmile Arm (survey 7)	15	0.085	107
Mean		0.100	135

between the interstitial waters, at about 135 μg/liter of total soluble P, and the water column, at about 3 μg/liter soluble P, is obviously quite steep. The sediments must therefore play some role in supplying phosphorus to the overlying water column, but there is no evidence that this internal source is unusual in Lake Dillon.

6. Particulates and Phytoplankton Biomass

The total particulate load of a water body can be divided into living and nonliving fractions. The living fraction is influenced by nutrients. The nonliving fraction, however, is jointly determined by biological factors such as the rate of production of biological detritus and by abiotic factors such as silt content of the lake's water supply. It is essential that the particulate functions be disentangled as well as possible before any causal analysis is attempted. Here we show the patterns of particulate concentrations in Lake Dillon and the contributions of identifiable fractions to these patterns.

Total Particulates in the Water Column

The average values for total dry weight of particulates retained on Whatman GF/C filters are summarized in Table 22. The table shows that the upper water column had a considerably higher total particulate concentration than the lower water column in both years. The averages for depths below 30 m were about half the averages for the surface. In the upper 25 m, total particulates were considerably higher in 1982 than in 1981 as an annual average. In both years the upper 5 m was slightly higher than any other depth, a layer from 5 to 15 m was distinctively high but not quite so high as the top 5 m, and the stratum from 15 to 25 m was a transition from the higher concentrations of the upper 15 m to the lower concentrations of the bottom 30 m of the water column.

Table 22. Mean Total Particulates of Lake Dillon in Different Strata of the Water Column (Index Station)

Stratum (m)	Particulates (mg/l)	
	1981	1982
0–5	2.23	3.85
5–10	2.12	3.59
10–15	2.14	3.64
15–20	1.77	2.84
20–25	1.59	2.31
25–30	1.44	1.62
30–35	1.23	1.57
35–40	1.19	1.27
40–45	1.42	1.98
45–50	1.24	1.52
50–55	1.66	1.40

Variations in total particulate concentrations with depth and between years are not really understandable without some detailed consideration of seasonal variation. Figures 27 and 28 show time–depth diagrams of total particulate concentration, from which a number of major seasonal variations are evident. Total particulate concentrations under ice were low at all depths. Inorganic particulates settle out very thoroughly under ice because of the lack of turbulence and very small amounts of incoming water.

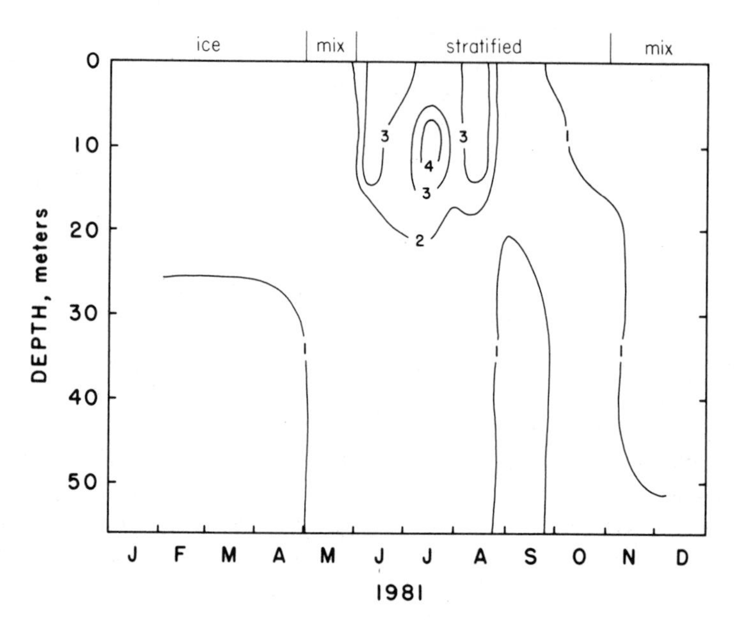

Figure 27. Time–depth diagram for total particulates in Lake Dillon (mg/liter), 1981.

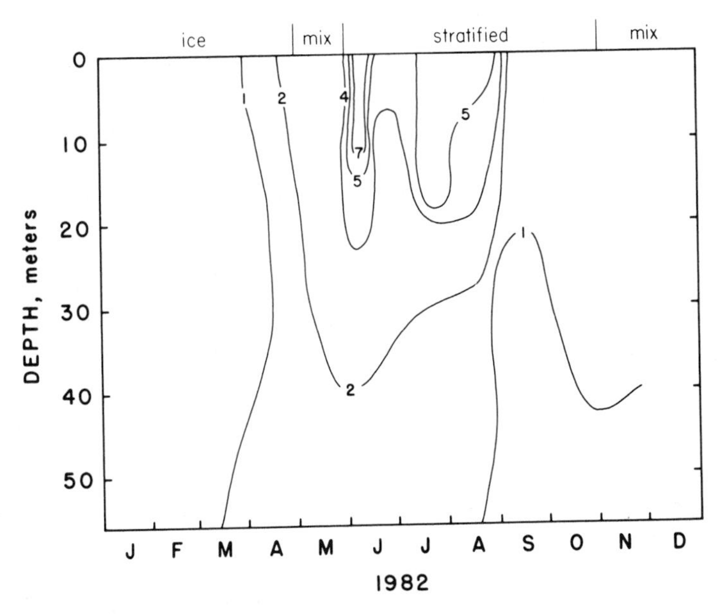

Figure 28. Time–depth diagram for total particulates in Lake Dillon (mg/liter), 1982.

Thus particulates that are present under ice are likely to be dominated by phytoplankton. Fine organic debris and bacteria may make small contributions, but microscopic examination of samples showed that these did not compare with phytoplankton in accounting for weight of filterable particulates under ice in Dillon. As already noted, the ice was not covered completely with snow in the first months of the 1981 ice cover. The conditions for development of phytoplankton in the upper water column under ice were thus exceptionally good. This is reflected in somewhat higher total particulate concentrations in the upper water column under ice in 1981.

During spring mixing, the winter phytoplankton were dispersed vertically and inorganic particulates entered from the watershed in increasing amounts. The stratification period, beginning in early June, was marked in both years by two peaks of particulate concentration occurring in the epilimnion. The first peak coincided with runoff, and was caused by the entry of large amounts of inorganic particulates from the watershed. As runoff declined, settling outstripped the addition of particulates. At the same time, as shown previously by Figure 19, incoming water from the rivers passed to greater depths as the summer progressed. Inorganic particulates entering deeper water would be expected to settle more quickly because of lower turbulence in deeper water. The peak for total particulates during runoff was considerably higher in 1982 than in 1981 because of the larger runoff in 1982. In both years there was a second peak of surface particulates in July after the peak runoff passed. The second peak of particulates was of similar magnitude to the earlier one associated with runoff. The second peak was caused by the buildup of phytoplankton biomass and not by inorganic particulates. This was confirmed by microscopic examination of samples and by chlorophyll analysis.

The amounts of total particulates in deep water during stratification were low by comparison with the surface values. This is principally explained by the smaller amount of turbulence in water below the thermocline, which would promote more rapid settling in deep water. In 1982 the total particulate levels below the thermocline were consistently greater than those in 1981, as would be expected from the entry and settling of greater amounts of inorganic particulate materials in 1982 with the greater runoff that year.

In both 1981 and 1982 there was a pronounced decline of total particulates in the surface water beginning in the last half of July. This was caused by a decline of phytoplankton biomass and will be discussed more fully below. Fall mixing established relatively uniform distributions of particulates.

Chlorophyll *a*

The distribution of chlorophyll *a* over time and depth is shown in Figures 29 and 30. There was a significant amount of chlorophyll under ice in both years, but the amount was much higher in 1981 than in 1982 because of the transparency of the ice and lack of snow cover in January and February of 1981. In fact a pronounced chlorophyll peak developed under ice in 1981, as shown in the time-depth diagram. The chlorophyll under ice in 1982 should probably be regarded as more typical, since the large amount of light reaching the water in January and February of 1981 was only possible because of the unusual weather conditions that year.

The spring mixing in May of 1981 drastically reduced the surface chlorophyll to a concentration well below 1 μg/liter. Two factors contributed to this decline. First,

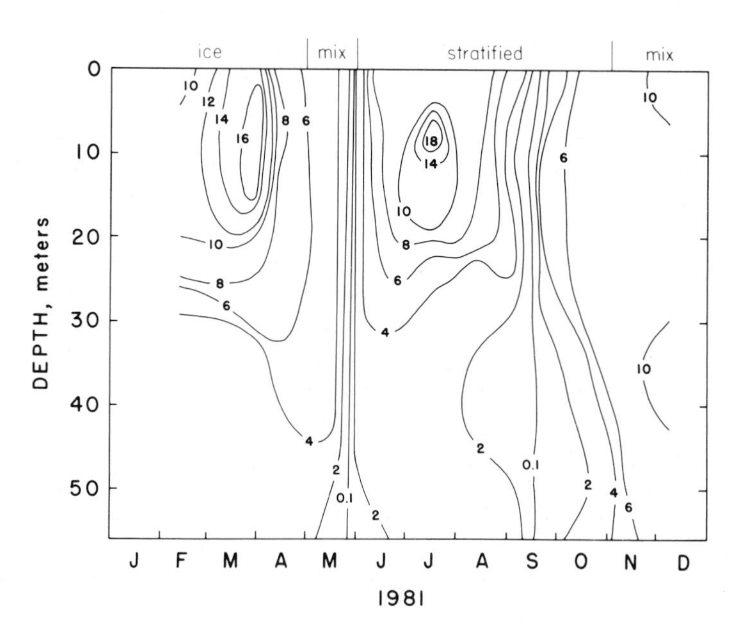

Figure 29. Time–depth diagram for chlorophyll *a* in Lake Dillon (μg/liter), 1981.

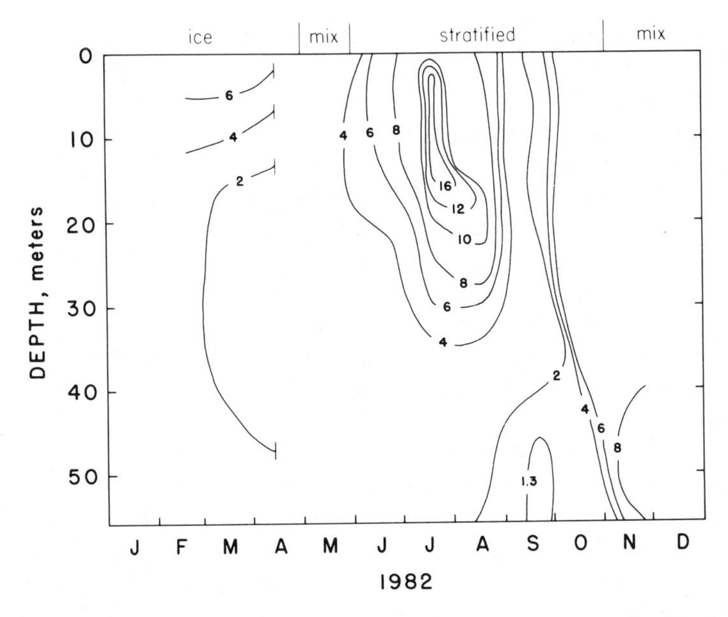

Figure 30. Time–depth diagram for chlorophyll *a* in Lake Dillon (μg/liter), 1982.

there was a reduction in surface chlorophyll that began well before the ice melted. This may have been caused either by light deprivation (due to light extinction by spring snow) or to nutrient depletion. An attempt to separate these effects will be made in Chapters 7 and 8. Second, the decline was accentuated suddenly at the time of melt by the mixing of surface water, already low in chlorophyll, with deeper water having essentially no chlorophyll. This was mainly a dilution effect. Dilution may have been accentuated by light deprivation due to deep mixing, which moved the algal cells to such great depths that they spent much time in the dark. This effect would have been short-lived, however, since incipient layering quickly began to impede mixing.

The light deprivation phenomenon can be shown more clearly by use of the concept of effective light climate (Ramberg 1976, 1979):

$$\bar{I} = I_o' \frac{(1 - e^{-\eta_t z_m})}{\eta_t \, z_m} ,$$

where $\bar{I}$ is effective light climate (mean exposure of the average cell in the mixed layer, expressed as langleys per day PAR), z_m is the thickness of the mixed layer (adjusted for morphometry), η_t is the extinction coefficient (m^{-1}), and I_o' is PAR at the top of the water column (langleys per day, with surface correction). In May, during mixing on Lake Dillon, z_m was about 25 m and η_t was about 0.5 m^{-1}. On a cloudy day I_o' would have been about 145 ly/day, and on a bright day it would have been about twice this amount. Thus for a cloudy day $\bar{I}$ would have been 11 ly/day and for a bright day it would have been 22 ly/day. The presumed threshold for phytoplankton growth is about 10–20 ly/day (Ahlgren 1979), so it is clear that really rapid growth could not have occurred until there had been a change in η_t or z_m. Spring changes in η_t were actually upward, reducing $\bar{I}$. Thus growth could only have occurred through a major change in z_m. Stratification caused the critical change in z_m, bringing $\bar{I}$ into the range 50-120 ly/day in early June.

The degree of decline in chlorophyll at the time of spring mixing in 1982 was not evident because samples could not be taken during the month of May, but in all probability it was very similar to the decline observed in 1981. In 1982, as in 1981, there was a rapid increase of chlorophyll concentration as soon as the water column stabilized in June. As is typical in dimictic lakes, the conditions for phytoplankton growth just after establishment of stratification in the spring were very good because available nutrient concentrations were at a maximum as a result of recent complete mixing of the water column, while daylength and water temperature were increasing and the phytoplankton cells were held within the lighted zone by the presence of a thermocline.

In both years the chlorophyll concentrations reached their peak in the first half of July. In 1981 the increase in chlorophyll was more or less uniform over the top 20 m until the first of July, and thus extended through the epilimnion into the metalimnion. Subsequent to the first of July, increases continued in the 5-10-m stratum, but not in the upper 5 m. Since the 5-10-m stratum incorporated part of the metalimnion, the chlorophyll maximum was at least partially metalimnetic. Metalimnetic maxima are common in stratified lakes (e.g., Fee 1976, Heaney and Talling 1980, Moss 1972, Moll and Stoermer 1982), since light often penetrates through the epilimnion and into the metalimnion in sufficient amounts to support phytoplankton growth. After nutrients are depleted from the epilimnion, growth may continue in the metalimnion where nutrients are still available, thus producing a metalimnetic maximum. The sunlight reaching Dillon on a clear day is about 50% greater than at the same latitude at sea level (Hutchinson 1957). This and the minimal cloud cover over the lake would promote metalimnetic growth by increasing the metalimnetic PAR.

In 1982 the summer chlorophyll peak coincided almost exactly with the timing of the 1981 peak, but the chlorophyll maximum was more evenly distributed over the top 15 m. As in 1981, phytoplankton were present in quantity well below the mixed layer, to a depth of about 20 m. Both the transparency data (Figure 16) and the primary production data (see below) indicate that growth was mainly confined to the upper 10 m, however, so the cells below this had probably grown in the upper 10 m and had sunk to greater depths. The more even vertical distribution over the top 15 m in 1982 is probably explained by the larger runoff and the tendency to runoff to enter higher in the water column in 1982. In 1982 significant amounts of nutrients continued to enter the mixed layer well after the first two weeks of June (Figures 12 and 19), and this may have retarded the occurrence of nutrient depletion in the uppermost portion of the water column.

The size of the chlorophyll maximum was very similar in the 2 years (Table 23). The maximum chlorophyll per unit volume was slightly higher in 1981 than in 1982, but the maximum chlorophyll at the lake surface (0-5 m) and the distribution of the maximum chlorophyll levels over depth were both greater in 1982 than in 1981. Thus the surface transparency of the lake was considerably less during July of 1982 than in July of 1981, despite the similarity in maximum chlorophyll concentrations of the 2 years. Because 1981 was a much more unusual year hydrologically than 1982, the 1982 chlorophyll data are probably more typical.

Table 23. Information on Chlorophyll in 1981 and 1982

	1981		1982	
	Amount	Time	Amount	Time
Maximum chlorophyll concentration on any day or depth (μg/l)	19.8	13 July	17.9	19 July
Maximum summer chlorophyll concentration on any day, surface (0–5 m, μg/l)	9.7	10 Aug	17.4	19 July
Maximum summer chlorophyll concentration monthly average, surface (0–5 m, μg/l)	9.3	June	11.7	July
Average chlorophyll, summer, 0–5 m (postrunoff stratification)	6.7	Jul–Oct	7.3	Jul–Oct

In both years a chlorophyll decline began in the middle of July. We attribute this decline to nutrient deficiency. In both years July was the month when available phosphorus and available nitrogen were first reduced to their summer minimum levels (Figures 22 and 24). The disappearance of critical macronutrients is only circumstantial evidence of the nutrient status of phytoplankton, however, because of the ability of phytoplankton to store significant amounts of macronutrients (Lund 1965). The hypothesis that the July decline in phytoplankton biomass was caused by nutrient stress will be further supported by other types of evidence, especially enrichment studies, to be presented in Chapter 8.

The decline of chlorophyll beginning in mid-July was different in the 2 years. In 1981 the decline of chlorophyll concentrations was steady at all depths. In 1982, however, the decline was steady in the top 10 m but there was a very deep maximum of chlorophyll centered at about 25 m in the first half of August. The transparency data for the same time interval show that significant photosynthesis cannot have occurred so deep in the water column (1% light at 10–15 m: Figure 16). We therefore conclude that the 25-m maximum represents sinking of viable but nongrowing photoplankton cells that formerly belonged to the surface maximum. In fact this sequence of events is suggested by the shape of the curves on the time-depth diagram for 1982 (Figure 30).

There was a steady decline in chlorophyll between late July and early September. This decline was caused by severe nutrient depletion, as will be shown in the analysis of production and enrichment studies. The thickening of the mixed layer in September renewed the nutrient supply, and thus led to an improvement in the nutritional condition of phytoplankton, as the enrichment data will show. The biomass of phytoplankton increased only slowly in response to nutrient renewal, however. Zooplankton analysis in Chapter 7 will show that zooplankton grazing peaked dramatically in September and was the probable cause of the slow response of phytoplankton to improved nutrition. Zooplankton declined after September, and phytoplankton then increased more rapidly.

The renewal of nutrients in mid-September was accompanied by declining light availability. Available light ($\overline{I}$, effective light climate) reached a minimum with complete mixing in November. In fact there was a definite slowing of the chlorophyll increase after October. If the entire water column were mixing day after day, $\bar{I}$ could not have exceeded 30 ly/day, which is near the presumed threshold of growth. Only strong winds will mix such a thick water column, however, especially in the daytime. Thus mixing can be sporadic even when water column stability is essentially zero. Interruption of mixing probably allowed rapid growth for short intervals in the upper 15 m. Algae in deeper water presumably did not grow but remained viable. The combination of growth in the upper water column and viability below under sporadic mixing allowed chlorophyll to increase even after the water column lost thermal structure, especially in 1982. Thus the Lake Dillon results suggest that sporadic stability may allow significant growth even when $\bar{I}$ is near the nominal minimum for growth.

Phytoplankton Composition and Seasonality

Table 24 summarizes the composition of the phytoplankton, and Figure 31 shows some of the more common taxa. Rare species are omitted from the table, and the remaining species are divided into categories according to their abundance. The dominant phytoplankton taxa are from three groups: blue-green algae, diatoms, and microflagellates.

By far the most important blue-green alga, and possibly the most important alga in the lake on an annual basis, is *Synechococcus lineare.* This species is often referred to as *Rhabdoderma lineare* in the literature, but has recently been renamed by Komarek (1976). The cells of *Synechococcus* are very small; in Lake Dillon, their diameter is very uniform at about 1 μm. The cell length is more variable, but averages about 15 μm. The cells are solitary or joined in pairs or units of 3 or 4, particularly during rapid growth, probably as a result of recent cell division. Because of its small size, this alga is frequently overlooked completely in quantitative counts, even when it is quite abundant. As a result, very little is known of its biology.

Lyngbya limnetica, the second blue-green that ranks as a dominant for Lake Dillon, was considerably less abundant than *Synechococcus* on an annual basis but did reach high abundance at certain times. This species of *Lyngbya* is characterized by extremely tiny elongate cells having dimensions and shape very similar to those of *Synechococcus.* The cells are united in filaments that are typically composed of 5 to 15 cells. The filaments are surrounded by a sheath, which is characteristic of the genus. This species is widespread, spanning the tropics to high north temperature latitudes, and is found in waters of widely divergent trophic status.

Neither of the two dominant blue-greens nor any of the other blue-greens listed in Table 24 is characterized by gas vacuole formation, nor is any of these species capable of nitrogen fixation. Large-celled vacuolate forms such as *Microcystis, Anabaena,* and large *Oscillatoria* were entirely absent in 1981 and 1982.

Three diatom species are classified as dominant in Table 24. *Asterionella formosa* is a large-celled colonial diatom common in the spring growth of temperate lakes in all parts of the world. *Synedra radians,* a solitary species, is also a common contributor to the diatom growth of lakes, and is distributed from the tropics throughout the temperate zone. Neither *Asterionella* nor *Synedra* is easily identified with any particular trophic state. *Rhizosolenia eriensis,* a major contributor to diatom biomass in 1981, is a much more unusual diatom dominant, although it is distributed in a wide variety of lakes at all latitudes. It may often be overlooked because of its extremely delicate frustule, which is almost invisible even under phase optics. The frustule is subject to dissolution even by relatively mild preservatives. In the Lake Dillon samples, the frustules disappeared within a period of 3 weeks in samples preserved with Lugol's solution. The presence of the diatom was detected during the examination of fresh samples and, once its presence was known, counts could be made at any later time from the protoplast, which is characteristic and resists dissolution.

Table 24. Summary of the Phytoplankton Composition of Lake Dillon, 1981–1982

Taxon	Abundance[a]
Cyanophyceae	
Synechococcus lineare	D
Aphanothece nidulans	C
Lyngbya limnetica	D
Lyngbya sp.	C
Chlorophyceae	
Monoraphidium contortum	C
Monoraphidium setiforme	S
Scenedesmus granulatus	C
Scenedesmus ecornis	S
Oocystis parva	C
Kirchneriella obesa	C
Cosmarium sp.	S
Coccomyxa	C
Chrysophyceae	
Dinobryon petiolatum	S
Dinobryon divergens	S
Bacillariophyceae	
Asterionella formosa	D
Nitzschia sp.	C
Synedra radians	D
Rhizosolenia eriensis	D
Stephanodiscus sp.	C
Cryptophyceae	
Rhodomonas minuta	C
Cryptomonas erosa	C
Unclassified	
Microflagellates	D

[a] D = dominant, i.e., ranking the top one or two species in abundance at certain times of the year; C = common, i.e., ranking in the top five species in abundance at some time of the year; S = secondary, i.e., reaching significant numbers but never ranking high in abundance.

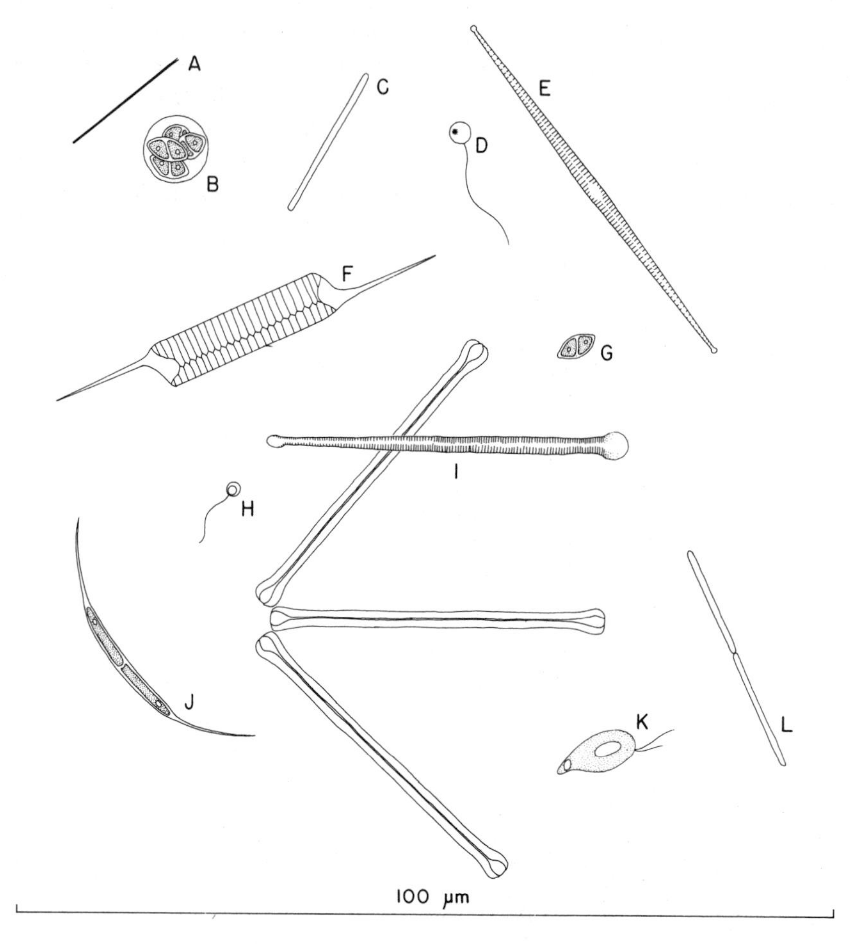

Figure 31. Scale drawings of Dillon phytoplankton: *Lyngbya* (A), *Oocystis* (B, G), *Synechococcus* (C, L), microflagellates (D, H), *Synedra* (E), *Rhizosolenia* (F), *Asterionella* (I), *Monoraphidium* (J), *Rhodomonas* (K).

The third group of species contributing to the dominants of Lake Dillon phytoplankton consists of unclassified microflagellates. These are extremely small Chrysophyceae and Chlorophyta that are generally recognized as important under certain conditions in temperate lakes but are virtually impossible to deal with taxonomically. The largest of these taxa in Dillon was about 8 μm in diameter and the smallest was just 1 μm in diameter.

The Chlorophyceae, Chrysophyceae, and Cryptophyceae were represented in the 1981–1982 Lake Dillon phytoplankton but did not contribute any dominants. Among the chlorophytes, *Monoraphidium* (formerly called *Ankistrodesmus*) (Legnerova 1969) was a steady contributor, as were the genera *Scenedesmus, Oocystis,* and *Kirchneriella.* Desmids were in general not well represented, although one unidentified *Cosmarium* species did make a significant appearance. A tiny spherical chlorophyte, *Coccomyxa,* was present at times. This is a widely distributed contributor to the so-called μ-algae (Lund 1961), which in Lake Dillon would also include the microflagellates. Two *Dinobryon* species, both common contributors to cold-water phytoplankton, were observed frequently but never became very common. The genera *Rhodomonas* and *Cryptomonas,* which are almost universal contributors to the phytoplankton of lakes and are extremely broad in their tolerance of trophic and thermal conditions, made significant contributions but never reached dominance.

It is typical of dimictic lakes to show major seasonal changes in composition of phytoplankton, and Lake Dillon is no exception in this respect. It is also common for dimictic lakes to show considerable variation in phytoplankton composition from one year to the next, especially if basic environmental conditions are markedly different. At the same time, there is typically a degree of repeatability in the seasonal sequence of composition from one year to the next (Lund 1971, 1979). Lake Dillon shows some major differences between 1981 and 1982, yet certain features were common to the 2 years. The years 1981 and 1982 probably approximate the maximum expected difference between 2 years in sequence because of the unusual hydrologic conditions in 1981 and the development of large amounts of phytoplankton under the ice in that year.

Figure 32 provides a simplified representation of the major changes in composition that occurred in 1981 and 1982 in Lake Dillon. The major chlorophyll peaks are shown from the chlorophyll time-depth diagrams, and the major contributors to these peaks are indicated. Significant but subdominant contributors are also indicated in parentheses.

The phytoplankton under ice in 1981 included large contributions by *Rhizosolenia,* whose average abundance in the top 25 m reached levels of about 16,000 cells/cc. The blue-green alga *Synechococcus* was also a major contributor numerically, although it is small and thus contributed less per cell to biomass (or to chlorophyll) than a number of other species (Table 25). *Synechococcus* reached abundances of about 6000 cells/cc under ice in 1981. Smaller but significant contributions were made by the chlorophyte *Kirchneriella* (about 1200 cells/cc) and the diatom *Asterionella,* whose peak numerical abundance was much lower (500-600 cells/cc), but which contributed substantially to the biomass because of larger cell size. Although *Asterionella* is principally a spring diatom, significant growth under ice is well known (e.g., Moss 1972).

The disappearance of ice cover in 1981 virtually eliminated the large *Rhizosolenia* population, but actually increased the *Synechococcus* population to about 12,000 cells/cc and brought up the *Asterionella* population to about 1700 cells/cc just at ice breakup.

The next major event in the phytoplankton dynamics of 1981 was the July peak, at which time the maximum annual chlorophyll levels were observed. In terms of cell numbers, *Synechococcus* was the dominant at this time (175,000 cells/cc). *Asterionella* also reached higher densities (about 3500 cells/cc). *Synedra* overtook *Asterionella,* however, with an abundance of about 9000 cells/cc. *Rhizosolenia* (about 8500 cells/cc) rebounded to about half its density under ice.

The July peak was followed by a general decline and by a more even distribution of abundance over depth as nutrient depletion became pronounced. As indicated in the analysis of chlorophyll data, there was a resurgence of phytoplankton following the thickening of the mixed layer in September. This resurgence was accounted for principally by increases in the abundance of *Synedra.*

In 1982 the large phytoplankton populations under ice did not appear as they did in 1981. The small populations that did develop under ice were dominated numerically by microflagellates, which reached abundances of about 2000 cells/cc, and by

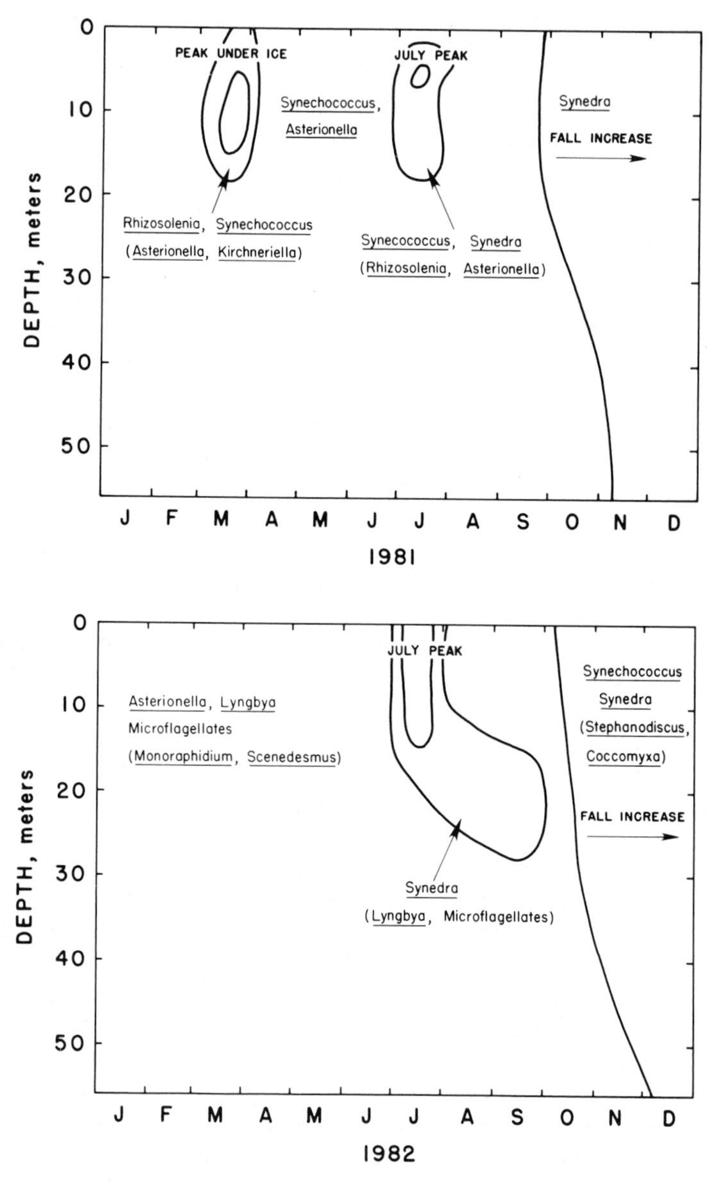

Figure 32. Simplified version of the chlorophyll time–depth diagram with the main contributions of algal taxa indicated. The most abundant taxa are shown without parentheses and taxa of secondary rank are shown in parentheses.

Table 25. Sizes of Important Lake Dillon Phytoplankton in Absolute Terms and Relative to *Synechococcus*

Taxon	Cell Volume (μm^3)	Ratio to *Synechococcus*
Synechococcus	12	1.0
Lyngbya	7	0.6
Asterionella	400	33.0
Synedra	300	25.0
Stephanodiscus	150	13.0
Rhizosolenia	25	2.1
Coccomyxa	0.2	0.017
Microflagellates	3	0.3

Lyngbya, which reached similar abundances. *Asterionella* reached abundances of about 1000 cells/cc, which was quite significant in terms of percent biomass because of the large cell size of *Asterionella.* The chlorophytes *Monoraphidium* and *Scenedesmus* also reached abundances of about 1000 cells/cc. The very large abundances of *Rhizosolenia* and *Synechococcus* that were observed under ice in 1981 did not make their appearance in 1982.

In 1982 the very large summer population of *Synechococcus* that was so evident in 1981 was not repeated, but the large summer growth of *Synedra* that appeared in 1981 was repeated in 1982 (18,000 cells/cc). Other diatoms made only minor contributions. Microflagellates and *Lyngbya* made contributions of about 4000 cells/cc each, and several species of chlorophytes made contributions of 1000 or 2000 cells/cc each. The fall growth was dominated by *Synedra* (6500 cells/cc) and *Synechococcus* (46,000 cells/cc), which appeared in large numbers after being rare throughout the earlier part of the year. Also appearing in the fall growth were large numbers of *Stephanodiscus* (3500 cells/cc) and *Coccomyxa* (12,000 cells/cc), neither of which had appeared prominently prior to fall 1982.

In summary, the stable elements of the annual composition cycle seem to be *Asterionella* growth under ice and in early spring and a major *Synedra* peak in July. Sporadic but repeated compositional elements include *Synechococcus* and *Rhizosolenia.*

Elemental Composition of Phytoplankton Biomass

Information is available on the amounts of carbon, phosphorus, nitrogen, and chlorophyll *a* in the particulate material (seston) of the top 5 m of Lake Dillon over the 2 years of study. Information of this type is sometimes useful in indicating the nutritional status of phytoplankton (Nalewajko and Lean 1980). There are a number of complicating factors that limit the usefulness of the data for nutritional interpretations in any particular situation, however. First, the seston is divided in an indeterminate way between living and nonliving fractions. The living fraction in Dillon is principally composed of phytoplankton (bacteria and protozoa are minor contributors on a weight basis). The nonliving fraction is composed of detrital organic matter and inorganic particles, principally silt or frustules from dead diatoms. When the phytoplankton contains a large contingent of diatoms, as is the case for Lake Dillon, the situation is further complicated by the presence of heavy and largely inorganic frustules around each living diatom cell. These frustules contribute 30 to 50% of the dry weight of diatom cells, depending on season and species composition (e.g, Bailey-Watts 1978). Microscopic examination of sedimented phytoplankton samples showed that Lake Dillon contained organic debris at virtually all times of the year, ranging from a minimum of 10-20% of total particulates under ice or at the height of stratification to 80-90% during spring mixing, spring runoff, and fall mixing. Thus the annual averages for element ratios would be virtually meaningless. It is worthwhile, however, to examine the element ratios for portions of the year when detritus contribution to filterable organic matter is known to be least. The appropriate times include the period of ice cover and the interval from the end of the first week in July, when runoff has ceased to affect surface water, to the middle of September.

The percentage of carbon in particulate materials filtered from the top 5 m of Lake Dillon varied from a sharp minimum of 5% at the peak of the 1982 runoff to a maximum of about 40% under ice during 1981 and 1982. The ratio of carbon to dry weight for phytoplankton, in the absence of diatom frustules, is about 36% (Stumm and Morgan 1981). Carbon percentages between 35 and 40% on four different occasions under ice coincided with the presence of large populations of nondiatomaceous algae, as expected from the literature values on percentage C. In all other seasons the percentage of carbon fell well below this, and was typically between 10 and 30%. In 1981, the percentage of carbon was typically higher than in 1982, probably because of the smaller relative contributions of diatoms in 1981 than in 1982. The sharp minimum of percentage carbon at runoff in 1982 is attributable to the inflow of significant amounts of silt with peak runoff. The coincidence of this low percentage carbon with low transparency confirms the interpretation that has already been given concerning the control of transparency by inorganic particulate material at this time of year.

Table 26 shows phosphorus as a percentage of organic matter (computed from carbon) for the two timespans during each year when the influence of nonliving material on the chemistry of seston was likely to have been lowest. Percentage of phosphorus in

Table 26. Composition of Organic Matter in the Top 5 m of Dillon at Times When the Contribution of Nonphytoplankton Sources to Organic Matter Was Lowest

	P as percent Organic Dry Weight	N as percent Organic Dry Weight	Chlorophyll *a* as percent Organic Dry Weight
1981			
Ice cover	1.0	13.8	0.87
Postrunoff stratification	0.8	8.1	0.38
1982			
Ice cover	1.4	–	0.73
Postrunoff stratification	0.6	4.9[a]	0.42

[a] With one unrealistically low value excluded.

phytoplankton protoplasm as reported in the literature generally falls between 0.5 and 3.0% of dry weight (Parsons et al. 1977), although lower concentrations are known and may sometimes be adequate to sustain high growth rates (Lund 1970). The observed ratios for Dillon were highest under ice, especially in 1982 when populations were small, and lowest during the postrunoff stratification. This is to be expected in view of the greater likelihood that nutrient depletion during summer stratification will be more extended and more severe than under ice cover.

Nitrogen percentages are also given in Table 26. The expected percentages are between 4 and 9% for healthy phytoplankton cells (Fogg et al. 1973). Three percent is considered the absolute minimum for phytoplankton growth (Fogg 1975). Under ice cover during 1981 the percentage was very high, perhaps unrealistically so, although phytoplankton are known to have significant storage capabilities for nitrogen (Fogg 1975) and are known to exceed 10% N on occasion (Lund 1965). Too many data are missing for the period of ice cover in 1982 to yield an estimate for that time. During postrunoff stratification both years the percentages were lower than under ice cover in 1981, and fell within the expected range for healthy cells. Significant changes in nitrogen stores may have occurred during the course of the postrunoff stratification, but these could not be deduced from individual determinations on field populations. Aside from the known presence of some nonliving material, resolution of changes over short time intervals is not really possible from individual determinations of element ratios because the degree of uncertainty for any given ratio determination.

Chlorophyll *a* can be expected to account for 0.5 to 2% of phytoplankton dry weight in most instances. The median is in the vicinity of 1% (Wolk 1973, Taguchi 1976, Gibson 1978, Bailey-Watts 1978). As shown in Table 26, the values for Dillon conform very well to these expectations. Under ice cover the values are near the midrange and in the postrunoff stratification the values are toward the bottom of the range of expected values (0.4%). Low ratios of chlorophyll to organic matter are symptomatic of nitrogen deficiency (Fogg et al. 1973). Nitrogen deficiency is more likely in postrunoff stratification than under ice, and this may explain the difference between the two seasons in the ratio of chlorophyll to carbon. Other factors, especially light adaptation, must also be considered, however.

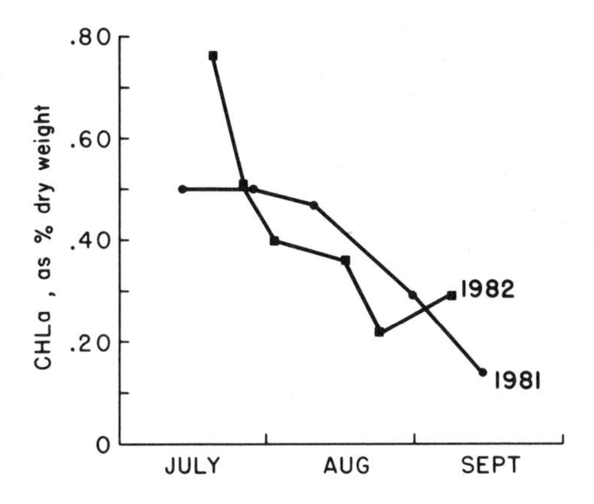

Figure 33. Chlorophyll *a* as a percentage of total particulate dry weight.

The degree of certainty that can be attached to the ratio of chlorophyll *a* to carbon is greater than for any of the other ratios, since chlorophyll *a* is not likely to be affected significantly by nonphytoplankton organic matter. For this reason, it is worthwhile to examine trends in individual values over the last half of the stratification season. Figure 33 shows the ratios of chlorophyll *a* to carbon between the middle of July and the middle of September for both years. In both years there was a trend toward lower ratios as time progressed. This downward trend is suggestive of nutrient stress, specifically nitrogen deprivation, and is part of the circumstantial evidence that concurs with the evidence of nitrogen deprivation provided by enrichment studies (Chapter 8).

7. Zooplankton

The zooplankton community of Lake Dillon is exceedingly simple in composition. More than 99% of the individuals in the community are accounted for by four species, including three species of rotifer and one species of copepod. The rotifers are *Polyarthra vulgaris* Carlin, *Keratella cochlearis* (Gosse), and *Keratella quadrata* (Müller). Each of these three species is found in a wide variety of lake types. Only the first two of these three rotifer species are really important, since *Keratella quadrata* does not reach abundances of more than 10% of either the other two species of rotifer. The Copepoda are represented by a single species (*Diacyclops bicuspidatus thomasi* Forbes). This is generally regarded as a cold-water species; at low elevations there are few reports of summer populations (Armitage and Tash 1967). Cladocerans (mainly *Bosmina*) appeared only in very small numbers during the study period.

Mysis relicta (Lovin) was introduced into Lake Dillon between 1969 and 1974, where it had established a good population by 1975 (Nelson 1981). Nelson's studies show that *Daphnia* populations, which had been substantial in earlier years, had declined to almost undetectable numbers ($\sim 1/m^3$) by 1978. It would appear that *Mysis* has essentially eliminated the cladoceran component of the open-water zooplankton in Lake Dillon, as it is known to have done elsewhere.

Figure 34 depicts the seasonal changes in the three major zooplankton species in 1981 (no counts were made in 1982). The rotifers *Polyarthra* and *Keratella* are shown in the top two panels. These species reached their minimum in late winter and began to show signs of increasing in number as soon as the ice cover was off. *Polyarthra* increased only slowly during the first half of the stratification season and fell to a

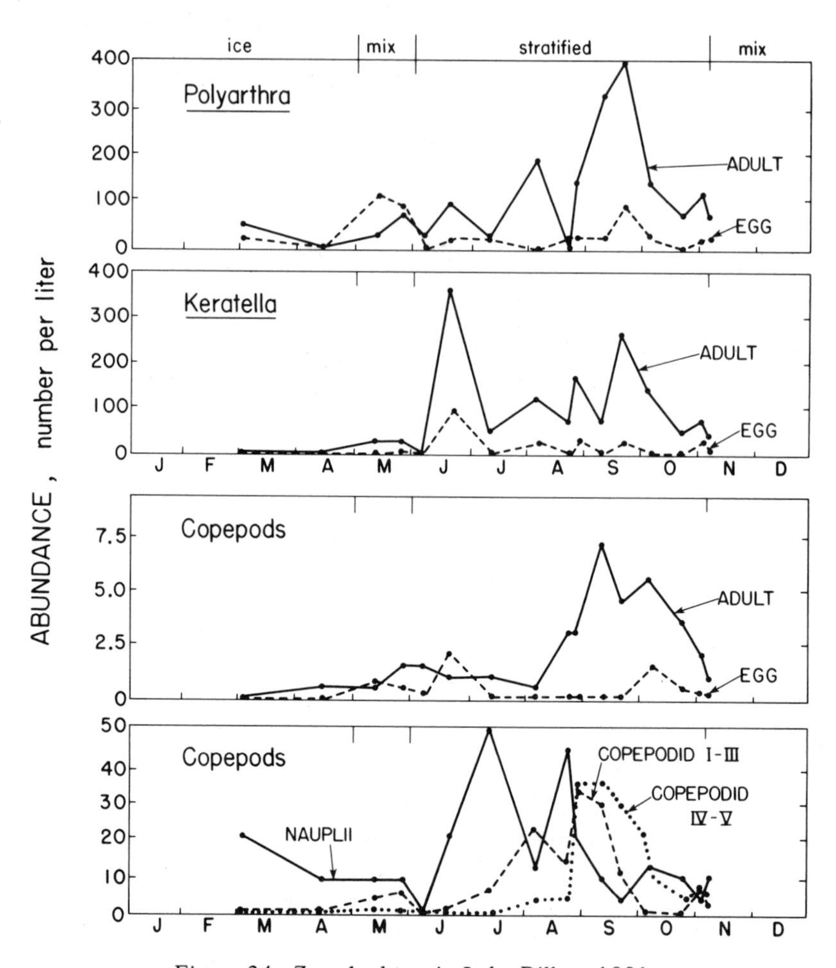

Figure 34. Zooplankton in Lake Dillon, 1981.

minimum at the end of August, when phytoplankton populations had declined and primary producers were under great nutrient stress. With the descent of the thermocline in September, which improved conditions for phytoplankton growth, *Polyarthra* showed a marked growth response which extended over a period of about 6 weeks. While there is little doubt that the fall increase of *Polyarthra* was connected with biological events triggered by the thickening of the thermocline in September, the exact mechanism causing increase remains mysterious, as is almost always the case with rotifer populations. Populations fell off slowly toward the end of the year.

Keratella showed a major peak just after the onset of summer stratification. This is frequently the case for this genus, which seems to be favored by conditions that coincide with or immediately follow mixing of the entire water column (cf. Lewis 1979, Makarewicz and Likens 1975). The population subsequently declined to intermediate abundances and showed a second peak after the descent of the thermocline in September in coincidence with the *Polyarthra* peak at that time.

At the peak population levels, and to some degree throughout the entire growing season, total rotifer populations were high. Rotifers were thus significant contributors to the total herbivore component. As shown by Makarewicz and Likens (1979) rotifers, although often treated as insignificant contributors to herbivory, can be important in a wide variety of lakes.

The bottom two panels of Figure 34 show the abundance of various stages in the life history of copepods in Lake Dillon during 1981. The adults and eggs are shown together in the uppermost panel. Adults were virtually absent under ice, but began to appear during the spring mixing. As soon as stratification was established in June, the adults produced a large number of eggs. The number of adults and the number of eggs thereafter declined to very low numbers until the individuals produced by the June egg pulse began to appear in the population as adults. This occurred at the end of August or the beginning of September. This new population of adults produced a fall pulse of eggs in October. Thus the population was bivoltine. At other locations, populations of this species are either univoltine (McQueen 1969, Cole 1955) or bivoltine (Le Blanc et al. 1981, Armitage and Tash 1967).

Significant numbers of nauplii were found under ice in the winter. This was the result of the fall egg pulse. A much larger group of nauplii appeared in June and July as a result of the egg pulse in June. In Figure 34 the June egg pulse can be seen moving not only through the naupliar stages but also into the first three copepodid stages and thence into the last two copepodid stages. Copepodids four and five from this cohort peaked in late August and early September. These copepodids were then the source of the surge in abundance of adults in September. Since the June cohort is clearly distinguishable from these graphs, it is possible to approximate the development time through the entire life cycle by comparing the timing of the peak for eggs with that of the peak for adults. The development time appears to have been almost 100 days from egg to egg. Some data are available on generation times of *Diacyclops bicuspidatus* from other lakes. The work by Le Blanc et al. (1981) and Armitage and Tash (1967) suggests generation times of about 2 months in the summer. Lake Dillon is considerably colder than the lakes they studied, which would help explain the longer development time for the Lake Dillon population.

The total copepod abundance was high only in September and October when the larger stages were present in greatest abundance. At other times total standing stock of copepods was relatively small. Although there is a diapause in some copepods (Hutchinson 1967), the Lake Dillon populations show no evidence of it. Pennak (1949), who studied copepods in seven Colorado lakes between 1600 and 2400 m, found no evidence of diapause.

Diacyclops bicuspidatus thomasi is an omnivore. Nauplii feed on phytoplankton (Neill and Peacock 1980), but the late copepodids and adults are carnivorous. Although the prey are typically smaller than the copepod (e.g., rotifers: McQueen 1969), there are records of adults eating larval fish (Smith and Kernehan 1981). Cannibalism is known (McQueen 1969).

The biomass of each of the species was approximated from the literature survey by Bottrell et al. (1976). The contributions of all species and developmental stages except the carnivorous C IV, C V, and adult copepods were then summed, and the results of this are depicted in Figure 35, which is useful as a guide to the probable importance of

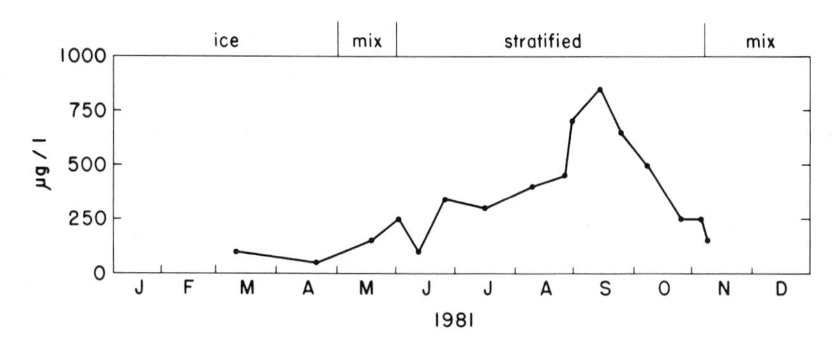

Figure 35. Total zooplankton biomass (wet), Lake Dillon, 1981.

zooplankton as herbivores in Lake Dillon. Figure 35 shows that the herbivore standing stock of zooplankton in Lake Dillon ranged from very low levels in winter to a level of a few hundred micrograms per liter wet weight in the first half of the summer and thence to a high of 700–800 μg/liter in September, after which there was a decline to previous levels. Although the feeding rates of zooplankton are not very well known and vary a great deal according to temperature and species composition (Gulati et al. 1982), a feeding rate for a mixed zooplankton assemblage would not ordinarily exceed 50–100% per day of the zooplankton standing stock (i.e., the daily ration is 50–100%). From Figure 29, which shows the chlorophyll in 1981, it is clear that the chlorophyll concentrations until the last half of August were generally in excess of 8 μg/liter. Since the ratio between chlorophyll *a* and wet phytoplankton biomass is approximately 1 to 1000 (see Chapter 6), there would have been in excess of 8000 μg/liter wet weight of phytoplankton. Thus making the liberal assumptions that the daily ration was 100% per day and that the grazers were relying entirely on phytoplankton for food, it is clear from Figure 35 that the zooplankton could not have been taking more than 10% per day of phytoplankton biomass prior to the end of August. Since the primary production data show growth rates well in excess of this amount, we conclude that zooplankton grazing cannot have been important in regulating phytoplankton biomass until the last part of August. In the last part of August, severe nutrient depletion reduced the growth rate substantially, yet zooplankton populations were increasing. When renewal of nutrients occurred due to thickening of the mixed layer in September, phytoplankton biomass almost disappeared, despite several consistent indicators that conditions for phytoplankton growth at this time were excellent. We conclude that zooplankton may at this time have played a role in the depression of phytoplankton biomass in the mixed layer. It is even possible that depression of the phytoplankton foods caused the decline of abundance for almost all zooplankton taxa subsequent to September. This in turn allowed the phytoplankton to build slowly in abundance, as observed in the chlorophyll and phytoplankton abundance diagrams.

In conclusion, it would appear that zooplankton are not important in controlling the total quantity of phytoplankton during most of the year, but assume importance as a source of phytoplankton mortality in the fall when high zooplankton populations coincide with phytoplankton abundances that have been depressed by poor nutritional conditions.

8. Photosynthesis and Oxygen Consumption in the Water Column

Photosynthesis

As described in Chapter 3, primary production measurements were made on Lake Dillon by the C-14 method. These data can be combined in several ways with information on chlorophyll and phytoplankton composition on the same dates. Four indexes related to primary production will be of interest here: primary production per unit area, column efficiency of photosynthesis, depth of the photosynthesis maximum, and maximum photosynthesis per unit chlorophyll.

Primary production per unit area (mg C/m^2/day) is affected mainly by the amount of light reaching the surface on a given day, by the amount of nonphytoplankton materials causing light extinction in the water column, by the size of phytoplankton populations, and by the physiological condition of phytoplankton populations (principally their internal nutrient inventories). Column efficiency of photosynthesis is the ratio of primary production per unit area to phytosynthetically available radiation (350–700 nm: PAR) penetrating the surface. We estimate PAR as 46% of total irradiance, and surface loss as 10% of PAR (Talling 1971). Since phytoplankton biomass and irradiance can both be expressed as energy, it would be possible to give the column efficiency as a true dimensionless efficiency number, but, since there is no particular advantage in this, the units to be used here are mg C/m^2/ly. This expression of photosynthesis is useful in that it removes most of the day-to-day variation caused by seasonal and irregular differences in daily insolation (e.g., Lewis 1974, Lewis and Weibezahn 1976). Main causes of variation in the column efficiency include extinction of light in the water column by nonphytoplankton materials, size of phytoplankton populations, and physiological condition of the phytoplankton (Bannister 1974a,b). The estimate of column efficiency is more subject to error during the period of ice cover than at any other time of the year, since the effect of snow on light penetration is difficult to quantify. For present purposes we have used an absorption coefficient for snow of 0.24 cm^{-1}.

Since phytoplankton populations are typically inhibited by high light intensities if they are held near the surface in the middle of the day, the photosynthetic maximum is usually found somewhat below the surface (Harris 1978). The depth of the maximum on a given day can be approximated from the shape of the photosynthesis profile on that day (Figure 36). The depth of the photosynthetic maximum is affected mainly by the amount of surface light, its extinction with depth, the vertical distribution of phytoplankton, and the light history of the phytoplankton (acclimation to high or low intensities). Under natural circumstances, cells in the mixed layer may not exhibit much inhibition because they are moved continually out of the inhibiting zone by water currents (Harris 1978). Although the inhibition segment of the curve may be exaggerated by confinement, it is useful in showing the seasonal changes in light response.

Photosynthesis per unit time at the depth of maximum photosynthesis can be divided by the amount of chlorophyll at the same depth, yielding a measure of the maximum photosynthesis per unit chlorophyll (mg C/mg chlorophyll *a*/hr), often represented as P_{max} · P_{max} is affected principally by the physiological condition of the phytoplankton and by temperature (Taguchi 1976, Harris 1978). The Q_{10} of P_{max} is about 2.0 (Harris 1978), so a 10°C rise in temperature will double P_{max} if other fac-

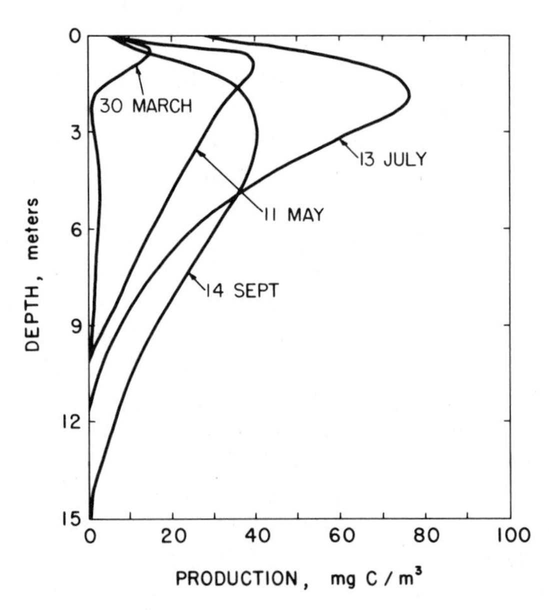

Figure 36. Production per unit volume versus depth for selected dates in 1982 as measured by C-14 uptake, showing shapes of photosynthesis curves.

tors are equal. If temperature is taken into account, P_{max} is the most direct indicator of the nutrient status of phytoplankton (Ganf 1975, Vollenweider 1965). P_{max} and the other three indexes for primary production are presented in Figures 37 and 38. Figure 39 shows P_{max} adjusted to a constant temperature of 15°C by use of Q_{10} = 2.0.

Production under ice during both winters was easily measurable, as would be expected from the chlorophyll levels. The absolute levels of production per unit surface averaged only about one-tenth the production levels observed in the summer, however. This is mainly explained by low light availability under ice, since the column efficiency was not low. Contrary to what would be expected from the much higher chlorophyll levels of 1981, the production under ice in 1981 was higher than in 1982 only on the first date of measure (23 February). This cannot be explained simply by the amounts of incident sunlight on the dates of incubation, as the column efficiencies of photosynthesis were similar under ice in 1981 and 1982. The higher chlorophyll levels of 1981 were thus not paralleled by higher production during the last half of the period of ice cover. Due to thin ice, no data are available for the first 2 months of ice cover, but it can be shown that high levels of chlorophyll under ice in 1981 must have been caused primarily by a burst of growth during January and February. The chloro-

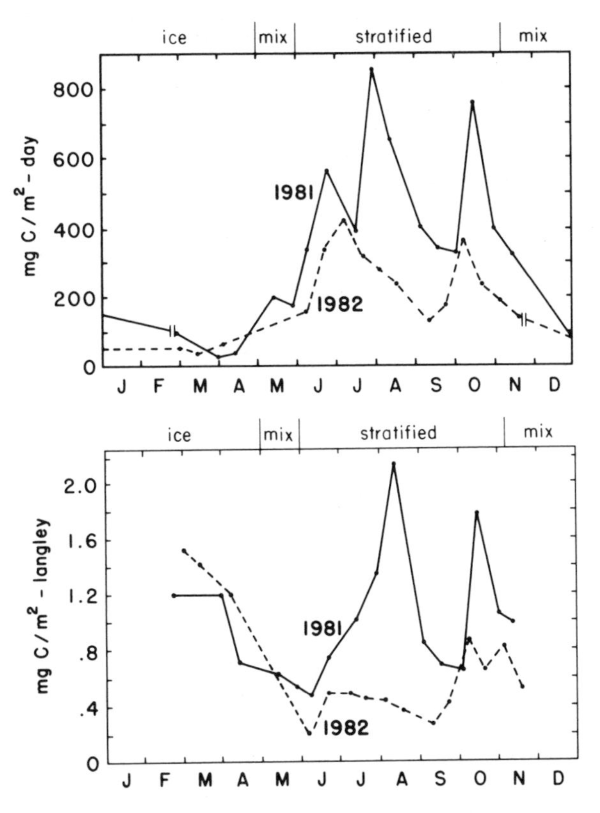

Figure 37. Primary production per unit area (above) and column efficiency of photosynthesis (below).

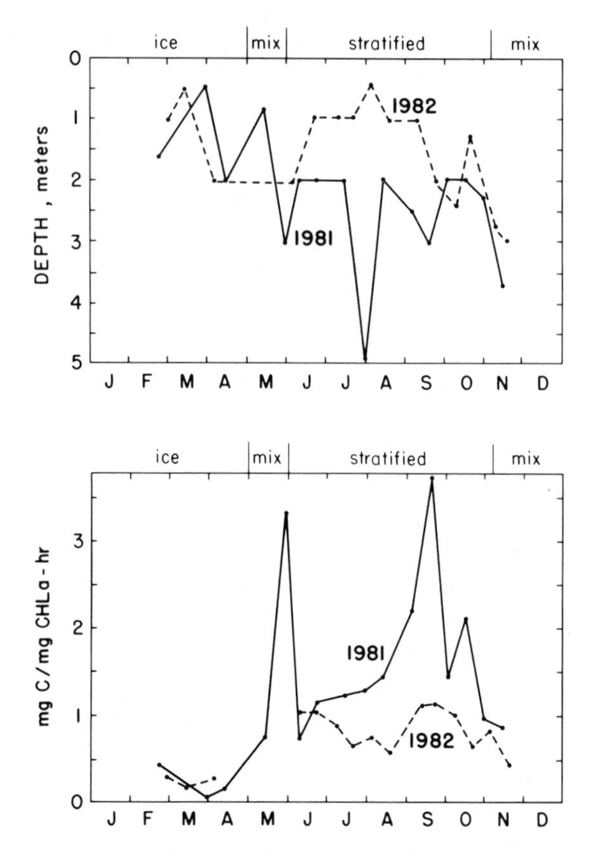

Figure 38. Depth of maximum photosynthesis (above) and maximum photosynthesis per unit chlorophyll (below).

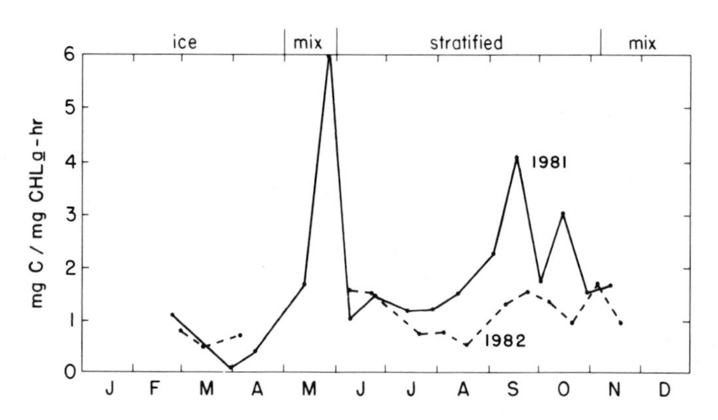

Figure 39. P_{max} corrected to a common temperature (15°C) by use of $Q_{10} = 2.0$.

phyll present at the time of first sampling thus would represent the accumulated biomass from this earlier growth, which was declining by the time the first sample was taken. Consistent with this explanation, the highest production figure under ice was the first one in 1981 (23 February). The total accumulation of chlorophyll between 1 January and the middle of March 1981 was at least 250 mg/m^2. From the ratio of chlorophyll to organic dry weight under ice in 1981 (Table 26), we estimate that this corresponded to 28 g of phytoplankton dry weight/m^2, or 11 g of carbon/m^2. Since the period of synthesis extended over 85 days, the average net synthesis per day must have been about 120 mg C/m^2. In view of the observed decline by early March, synthesis during January and February must have exceeded 120 mg C/m^2/day.

The P_{max} values were low under ice both years, even after temperature adjustment (Figure 39), suggesting nutrient limitation. Nutrient limitation would almost certainly have been due to low amounts of available P, since N was available in quantity under ice (Figures 25 and 26). When light is available in such low amounts, however, P_{max} must be interpreted with caution, since light may be insufficient to saturate photosynthesis. The consistency of surface inhibition (Figure 38) indicates that saturating intensities were reached during all incubations, however, despite the low surface intensities in winter. Saturation at low intensities is not unusual for phytoplankton adapted to low light. Saturation at intensities as low as 30 μEinst/m^2/sec are known (Harris 1978), and the Dillon intensities exceeded this even under snow. We therefore conclude that the P_{max} values were truly low under ice, and are thus indicative of P depletion.

During the spring mixing, production per unit area was intermediate between production of the period of ice cover and production of the period of stratification. Column efficiency of photosynthesis was also intermediate, but P_{max} was high. Column efficiency was suppressed by extreme vertical dispersion of the chlorophyll, which caused most of the light to be taken up by nonchlorophyll absorbance (see Figure 17). At the same time, P_{max} was high because the phytoplankton, although too dispersed to harvest the light efficiently, were in top physiological condition as a result of nutrient redistribution by mixing.

Production per unit area rose shortly after the onset of stratification in both 1981 and 1982. This is explained by the increasing amounts of light due to longer days, by the larger stock of chlorophyll, and by the reduced vertical dispersion of chlorophyll after the cessation of mixing. Column efficiency, however, decreased to a minimum in early June of both years. This was a result of runoff, which added large amounts of particulates capable of taking up most of the light before it reached phytoplankton chlorophyll. The effect was much more prolonged in 1982 because of higher runoff and shallower entry of runoff into the water column that year. P_{max} in 1981 declined after the onset of stratification from its spring peak, suggesting the onset of some nutrient limitation very shortly after stratification began. In 1981 the June P_{max} values were of similar levels, although the decline was not documented because the lake could not be sampled in May.

In 1981, photosynthesis per unit area and column efficiency climbed steadily until late July or early August, after which there was a precipitous decline. The situation was very different in 1982. The rise of both variables, which began much the same way as in 1981, leveled off in June and began a slow decline in July. The depth of maximum fixation was much closer to the surface than in 1981 and P_{max} declined in parallel with total production. Two factors explain the difference between years: nutrient depletion and interception of light by inorganic particulates. Nutrient limitation began early in the stratification period of both years, as shown by the decline in P_{max} from spring mixing. In 1981, inorganic particulate loading was small after June and passed into the water column below the producing zone. Low chlorophyll at the surface, caused by nutrient depletion there, allowed the chlorophyll peak to develop deeper in the water column where nutrients were more available. This accounts for the increasing depth of maximum fixation. In 1982, production in deeper water was not possible because of lower transparency caused by much higher particulate input to the lake, and by the proximity of this input to the surface. Surface nutrient depletion was thus not compensated by downward shift in the depth of maximum production in 1982, and production was consequently lower.

In both years there was a major thickening of the epilimnion in mid-September. The immediate consequence of this was a rise in P_{max} caused by the relief of nutrient depletion. Production per unit area and column efficiency were slower to respond, since there was a dilution effect and grazing suppression along with the improvement of nutrient supply. By the end of September, however, these variables also showed a peak. Biomass buildup was accompanied by a return of lower P_{max} and a downturn of production in October. A portion of the autumn decline in P_{max} is accounted for by declining temperature (Figure 39). Decline in the temperature-corrected values suggests some nutrient limitation also, however. Since nitrate was abundant at this time, phosphorus depletion was the probable cause of decline in temperature-corrected P_{max}.

Among well-studied lakes exposed to similar climatic conditions, Lake Dillon has an annual production similar to that of Lake Erken (Sweden), which is considered mesotrophic (Nauwerck 1963). In general, oligotrophic lakes produce less than 300 mg C/m^2/day during the growing season, whereas mesotrophic lakes produce 300 to 1000 mg C/m^2/day (Likens 1975). Since the average production of Lake Dillon was well above 300 mg C/m^2/day in 1981 and near 300 mg C/m^2/day in 1982, the lake would be most reasonably classified as mesotrophic on the basis of production.

The P_{max} values of Dillon are comparable to those observed in other lakes. For example, Glooschenko (1973) found that P_{max} in Lake Huron varied between 0.5 and 3.5 in the ice-free season; this range is similar to Dillon's.

The pattern of production was essentially the same in 1981 and 1982: spring maximum, early fall maximum, summer depression, and late fall depression. This is a common pattern for temperate lakes (diacmic: Hutchinson 1967). Despite the similarity in pattern, the 2 years differed considerably in total production. The lower production in 1982 was caused by earlier termination of the spring rise in production, which has already been explained in terms of nutrient depletion and inorganic turbidity.

Table 27 compares the primary production of Lake Dillon with that of other lakes of the Colorado Rockies. The comparisons cannot be considered definitive, since complete annual data are available only for Dillon. The comparisons are based on single measurements for a given day within 2 days of 31 August on the year indicated in the table. Since the comparative data were taken in the mid-1960s, some of the produc-

Table 27. Comparison of Primary Production, as Measured by C-14, between Dillon and Other Colorado Lakes[a]

Lake	Year	mg C/m^2/day
Green Mountain	1963	108
Carter	1964	153
Dillon	1982	279
Granby	1963	351
Grand	1963	387
Dillon	1981	402
Shadow Mountain	1963	459
Estes	1964	468
Horsetooth	1964	774

[a] All production figures are for single measurements made within 2 days of 31 August in the indicated year. All data except for Dillon are from Nelson (1971).

tion values may now be higher at the same time of year as a result of additional nutrient loading. The table indicates that the production of Dillon, at least in the early fall, is within the midrange of values to be expected in lakes and reservoirs of comparable size.

Oxygen Concentrations and Oxygen Consumption

Oxygen depletion is of direct interest because of its significance to bottom fauna and deep-water organisms, including fishes, and because of the major change in internal nutrient flux that occurs when the oxidized microzone at the mud surface becomes reduced (Hutchinson 1957, Golterman 1976). The deep-water oxygen concentrations of a lake are also of indirect interest as an indicator of trophic status, provided that correct compensation is made for morphometric variation.

The oxygen concentrations of Lake Dillon in 1981 and 1982 are shown in the time–depth diagrams of Figures 40 and 41. Table 28 shows the oxygen concentrations corresponding to saturation at the elevation of Lake Dillon at various temperatures. It is evident from the table and from the time–depth diagrams that the surface waters of Lake Dillon were near saturation throughout 1981–1982. Production was never high enough to hold oxygen concentrations much above saturation. The most extended surface deviations from saturation occurred under ice, when atmospheric exchange was impeded. Surface oxygen concentration was slightly above saturation under ice in February of 1981. This was a result of the unusual winter burst of production in January and February, when the ice was clear and free of snow. Other instances of supersaturation were very minor. There were also a few instances of subsaturation near the surface. A significant depression of surface oxygen concentration occurred in September 1981. This was caused by thickening of the mixed layer and incorporation of deeper, less oxygenated water with surface water. September suppression of surface oxygen was also detectable in 1982, but was considerably smaller because the speed of

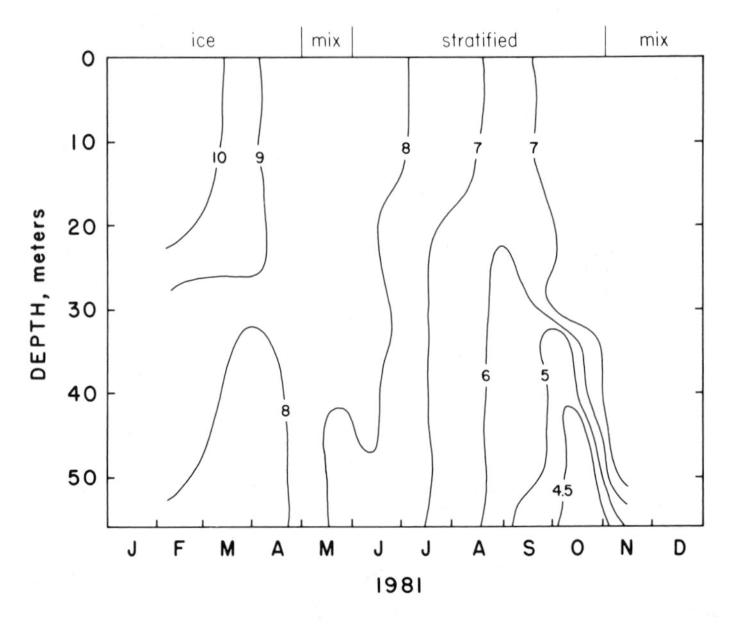

Figure 40. Time–depth diagram for oxygen in Lake Dillon, 1981 (mg/liter).

incorporation of deeper water was slower in 1982 than in 1981. Complete mixing in the first week of November also lowered the oxygen concentration at the surface, but not enough to be evident in Figures 40 and 41.

Dissolved oxygen concentrations in deep water did not remain near saturation throughout the year. Deep water became saturated during spring and fall mixing. The

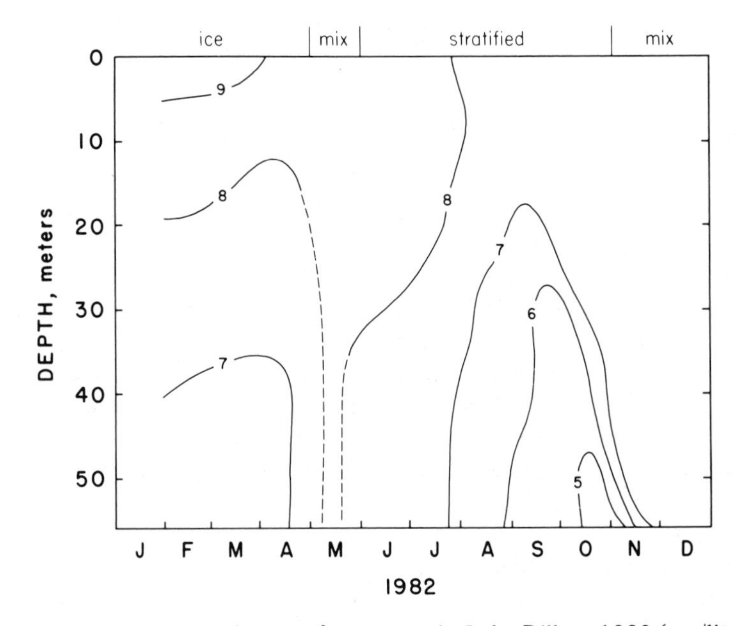

Figure 41. Time–depth diagram for oxygen in Lake Dillon, 1982 (mg/liter).

Table 28. Percent Saturation Corresponding to Various Temperatures and Oxygen Concentrations at the Elevation of Lake Dillon

Temperature (°C)	Oxygen (ppm)					
	5	6	7	8	9	10
0	50	59	69	79	89	99
5	57	68	79	91	102	113
10	64	77	90	103	116	128
15	72	86	101	115	129	144

water column thus entered the period of winter ice cover or the period of summer stratification with no oxygen deficit. There was significant depletion of oxygen in deep water both under ice and during summer stratification, however. The bottom 10 m of the water column under ice declined to a winter minimum between 6 and 7 ppm, as compared with a saturation concentration of about 9 ppm at the prevailing temperatures. Deep-water oxygen depletion during summer stratification was considerably more severe, partly because of the greater length of time over which depletion could occur and partly because of more rapid supply of organic materials subject to decay during the summer months. Oxygen concentrations 5 m above the bottom declined to 4.4 ppm in 1981 and to 4.6 ppm in 1982. Perhaps even more significant than the minimum concentration in the deepest water was the great vertical extent over which notable oxygen depletion occurred. The bottom half of the water column was markedly affected in both years. Since deep-water oxygen concentrations at saturation would have been about 9 ppm, the observed oxygen levels were as low as 50% of saturation.

There have been a number of efforts to link oxygen depletion quantitatively to the productivity of lakes. Although it is generally accepted that the linkage exists, the quantitative complications are numerous. First and most obvious is the dependence of the degree of oxygen depletion on the volume of oxygen-rich water at the beginning of the season during which oxygen depletion occurs. This so-called "morphometric effect" (Hutchinson 1957) is dependent on the ratio of the epilimnion volume to hypolimnion volume in a given lake. To compensate for the different ratios of layer volumes in various lakes, and thus achieve a comparison between lakes, the concept of areal hypolimnetic oxygen deficit (AHOD) was introduced (Strom 1931, Hutchinson 1938). The average amount of oxygen in the hypolimnion per unit surface of hypolimnion is computed by integration of oxygen concentration with depth from the top of the hypolimnion to the lake bottom. The amount of oxygen per unit area is then plotted against time, and the slope of the decline is determined by linear regression (Lasenby 1975). The slope is the AHOD, expressed as milligrams of oxygen per square meter per day (mg O_2/m^2/day). The terminology, which is dictated by traditional usage, is somewhat confusing in that AHOD is not really a deficit, but rather a depletion rate.

Among different lakes, AHOD is significantly correlated with trophic indicators such as total phosphorus, primary production, and transparency (Cornett and Rigler 1980). These relationships are potentially useful in predicting the degree of oxygen deficiency to be expected in a lake in response to changes in trophic status. However,

Table 29. Information Required for Prediction of AHOD by the Equations of Cornett and Rigler (1980)

Year	Mean Total P (0–5 m, μg/l)	Mean Summer Secchi Depth (m) (25 June–30 Sept)	Primary Production ($g\ C/m^2/yr$)	Mean Depth (m)	Mean Hypolimnetic Depth (m)	Mean Hypolimnetic Temp. (°C)
1981	7.63	3.14	110	24.1	23.7	8.9
1982	9.15	2.78	57	24.1	23.7	6.9

despite the corrections for major differences in layer volumes and in the duration of stagnation that are inherent in the computation of AHOD, there is a significant amount of scatter in all of the relationships between trophic indicators and AHOD. Some scatter is removed by use of depth in the equations, but as yet there seems to be no simple method for reducing scatter beyond this (Cornett and Rigler 1980).

The AHOD was calculated for Lake Dillon over the 1981 and 1982 summer stagnations. The decline of hypolimnetic O_2 per unit area was very linear both years, thus the AHOD could be estimated within narrow limits. In 1981 the AHOD was 710 mg O_2/m^2/day (standard error, 40) and in 1982 it was 630 mg O_2/m^2/day (standard error, 37). These are considerably higher than the values for oligotrophic lakes given by Lasenby (1975).

Three separate relationships were developed by Cornett and Rigler (1980) to predict AHOD from key trophic indicators. Table 29 shows the information required to make the predictions and Table 30 shows the predicted and observed AHOD values for Dillon. The phosphorus equation substantially underestimates AHOD for Dillon. Dillon produces about double the oxygen deficit that would be expected from P concentrations. This is due to the higher biological activity supported by a given P concentration in Lake Dillon than in most lakes; this phenomenon will be documented and explained in chapters to follow. The equation based on primary production is also a

Table 30. Observed Areal Hypolimnetic Oxygen Deficits in 1981 and 1982 Compared with Those Predicted from Three Separate Equations Developed by Cornett and Rigler (1980)

		Predicted AHOD ($mg\ O_2/m^2/day$) Using Equations with Indicated Variables[a]		
Year	Observed AHOD ($mg\ O_2/m^2/day$)	$P, \bar{z}$	Secchi, $\bar{z}$	Production, $\bar{z}_H$, T_H
1981	710	295	580	4841
1982	630	323	636	636

[a] P, mean total P concentration; $\bar{z}$, mean depth; $\bar{z}_H$, mean hypolimnion depth; T_H, volume-weighted mean temperature of hypolimnion.

poor predictor, at least for 1981 when hypolimnetic temperatures were exceptionally warm. The poor performance of the production equation is not surprising, since Cornett and Rigler found it to be the least reliable of the equations they tested. Secchi depth produces the best predictions. The AHOD for 1981 is underestimated by about 20% and the AHOD for 1982 is estimated almost exactly.

9. Nutrient Enrichment Studies

The purpose of a nutrient enrichment study is to determine the actual or incipient limitation of phytoplankton by specific nutrients. The literature on nutrient enrichment contains a wide variety of experimental designs and methods for such studies. This variation in methodology is in part explained by differences in the exact purpose of the studies. In general, nutrient enrichment studies can involve enrichment of standard algal cultures, or enrichment of natural assemblages over a few hours, a few days, or many days.

For the enrichment of laboratory algal cultures with lake water filtrate, a water sample from the lake is passed through a filter sufficiently fine to remove the phytoplankton. The water is then added to standard algal cultures, along with a spike of phosphorus, nitrogen, or any other individual nutrient or nutrient combination that might be of interest (e.g. Maloney et al. 1972). The response of the standard alga (often *Selenastrum capricornutum* in the United States) is measured in terms of cell number or chlorophyll. The responses in comparison to the control (no nutrient additions) are taken as an indication of the identity of the potentially limiting nutrient in the lake. The special advantages of this method are that it does not require extended field work and that it provides a common basis for the comparison of very different kinds of lakes. The disadvantage is that standard laboratory cultures differ considerably from a natural algal assemblage. For example, it is well known that algal cultures, either through physiological acclimation or genetic adaptation, typically require much higher amounts of nutrients to produce significant growth than do field populations (Nalewajko and Lean 1980, Lund 1965).

Short-term enrichment of the natural community (e.g., Goldman 1972) is probably the most common type of enrichment study when the emphasis is on the identity of actual (not potential) limiting nutrients at a given time in a particular lake. A sample is taken from the growth zone and placed in a number of bottles. Nothing is added to certain bottles; these serve as the control. Other bottles are enriched with individual nutrients or nutrient mixtures. All bottles are then inoculated with C-14 and resuspended in the euphotic zone or placed in an incubator. Production is estimated from C-14 uptake over a period of a few hours. In theory, algae that are limited by a given nutrient will show increased photosynthesis in response to the addition of that nutrient. Despite the extensive use of this method, a critical flaw in the underlying assumptions has recently been emphasized by Healey (1979) and Lean and Pick (1981), and has been well demonstrated numerous times in laboratory cultures (Rhee 1980, Eppley 1981). The response of nutrient-starved cells for the first few hours or more after their exposure to new supplies of the critical nutrient is diversion of cellular resources to uptake of the nutrient rather than to growth. A few hours may therefore be simply too short a time in which to measure the growth response of phytoplankton to the addition of a limiting nutrient.

Enrichments of the natural community lasting a few days are identical to the short-term enrichments except that the nutrient treatments are suspended in the lake for several days instead of a few hours (e.g., Lewis 1983b). The growth response is typically measured by change in chlorophyll with respect to a control rather than by C-14, as the C-14 method is not valid for application over extended time intervals. The main disadvantage to this method is the artificiality of enclosure conditions as compared with open-water conditions. This disadvantage can be combatted by various means, including the use of large containers with soft sides rather than small bottles.

For long-term enrichments of the natural community (e.g., Gerhart and Likens 1975, Lund 1981), large corrals or enclosures are used. Although this is an informative method, it typically is too expensive for routine application. Furthermore, the information it yields is more relevant to the projection of community changes in response to enrichment than it is to the instantaneous nutritional condition of a community.

Lake Dillon was studied in 1981 and 1982 by enrichments of the natural community lasting a few days. All enrichments were carried out at the index station. On the day of each enrichment study, integrated samples of the top 5 m of the water column were removed from the lake and placed in a large chamber where they could be homogenized. These were then siphoned into a number of soft plastic containers, each with a volume of 10 liters. Certain of the containers were left unaltered as a control. Other containers were enriched with KH_2PO_4 in amounts sufficient to increase the original phosphorus by 100 μg/liter in the container. Nitrate enrichment was by means of KNO_3 to a concentration of 200 μg/liter. The amount of enrichment was intended to exceed any possible requirement of either N or P over the interval of incubation but to be below the amount that might have toxic effects. The number of replicate containers for each treatment and for the control was typically three, and occasionally four. All except one enrichment study involved a nitrogen treatment, a phosphorus treatment, a phosphorus plus nitrogen treatment, and a control. The first enrichment study lacked the nitrogen treatment.

After the containers were filled and spiked as needed, they were all resuspended at a depth of approximately 2 m in the water column. This depth corresponds very closely to the expected depth of maximum photosynthesis (Figure 38). Because the containers were hooked to a floating buoy, they moved up and down with the waves at the lake surface. This and the incorporation of an air bubble inside each of the containers induced sufficient turbulence to keep the algae inside the containers in suspension. The containers were left in place for a period of 3 to 7 days, depending on the time of year. When the containers were removed from the lake, they were taken to the Snake River Wastewater Treatment Plant, where water from each container was filtered for chlorophyll analysis as described in Chapter 3. Some analyses were also done for other variables.

Three enrichments were done in 1981. In 1982 the number of enrichments was increased to 7 because the 1981 data indicated a shift in the limiting nutrient, and the timing of the shift could only be identified with certainty on the basis of more closely spaced enrichments.

The chlorophyll concentrations for each enrichment were analyzed by one-way ANOVA in which the enrichment treatment served as the variable of classification and the dependent variable was the final chlorophyll concentration. The null hypothesis was that different enrichments were not significantly different in chlorophyll from each other or from the control.

Table 31 summarizes the results of the enrichment studies. The one-way ANOVA showed that the mean response to the four treatments was not homogeneous on any one of the 10 dates ($p<0.01$). In other words, the four treatments could be divided into two or more groups on any one of the 10 dates. In order to determine which of the treatments on any given date were statistically distinct from each other, the Student–Newman–Keuls (SNK) multiple range test was applied to the data. Table 32 separates the four treatments into statistically distinct groups according to the SNK test. The groups are arranged in the table in order of ascending response to enrichment.

Table 31. Chlorophyll Concentrations (μg/l) for the 10 Enrichment Studies

Date	Treatment[a]				Dominant Limitation	Ratio (P + N/C)
	C	P	N	P + N		
1981						
20 July	4.4	5.7	–	20.5	N	4.7
24 August	2.1	2.2	5.7	20.6	N	9.8
28 September	5.3	5.8	5.0	11.1	P/N	2.1
1982						
24 May	2.7	3.9	3.2	3.8	P	1.4
14 June	6.8	11.7	6.8	11.1	P	1.6
10 July	8.3	17.1	8.9	20.5	P	2.5
26 July	7.9	7.1	11.9	39.0	N	4.9
23 August	5.0	4.9	12.0	25.4	N	5.1
7 September	1.8	2.0	2.3	2.6	P/N	1.4
18 October	6.5	13.4	6.8	14.5	P	2.2

[a] C, control; P, phosphorus only; N, nitrogen only; P + N, phosphorus plus nitrogen.

Table 32. Enrichment Treatments Separated into Statistically Coherent Groups Based on the Student–Newman–Keuls Multiple Range Test ($p<0.05$)[a]

Date	Group 1[b] (Lowest)	Group 2	Group 3 (Highest)
1981			
20 July	C, P	P + N	–
24 August	C, P, N	P + N	–
28 September	C, P, N	P + N	–
1982			
24 May	C, N	P, N, P + N	–
14 June	C, N	P, P + N	–
10 July	C, N	P	P + N
26 July	C, P	N	P + N
23 August	C, P	N	P + N
7 September	C, P	P, N	N, P + N
18 October	C, N	P, N + P	–

[a] Groups are arranged from left to right in order of increasing chlorophyll response. A given treatment can belong to more than one group if it is statistically inseparable from more than one group.
[b] C, control; N, nitrogen only; P, phosphorus only; P + N, phosphorus plus nitrogen.

Since there was always a statistically significant heterogeneity among the four treatments, it is defensible to identify a limiting nutrient for each of the 10 dates. This is accomplished by the combined use of Tables 31 and 32. The limiting nutrient is here operationally defined as the nutrient producing greatest growth response. Multiplicative effects may exist by which one nutrient interacts with another (Droop 1973), but the present weight of evidence seems to favor a sharp transition between P and N limitation for individual species (Rhee 1980). In a community, responses to two different enrichments could represent different limitations among the species.

The degree of response to enrichment, which can be expressed conveniently as the ratio of the phosphorus plus nitrogen treatment to the control, varied considerably between dates, as indicated in the last column of Table 31. Certain features were common to all the enrichment studies. The P + N treatment always fell in the highest group of responses, as shown in Table 31. In 6 of the 10 enrichments, the response to P + N was higher than the response to either nutrient alone. Given the relatively short duration of the incubations, such a result suggests that phosphorus and nitrogen requirements are closely balanced to the supply of these nutrients, even though one or the other may be identifiable at a particular moment as the dominating control on phytoplankton growth. Addition of either nutrient in quantity relieves the deficiency of that nutrient but induces deficiency of the other; thus the highest response occurs when both are added. Another feature common to the enrichment studies is the appearance of the control (i.e., no enrichment) in the lowest statistical response group. Since this is expected given the assumptions of the studies, it suggests that the studies were methodologically sound.

The number of enrichments was much smaller in 1981 than in 1982. Where the enrichment studies overlapped for the 2 years, however, the results are so similar as to suggest a common pattern. In July and August of 1981 the response to phosphorus plus nitrogen was very strong but the response to phosphorus only was very low or nil. Unfortunately, a nitrogen-only enrichment is not available for 20 July 1981, but the low P response and the high P + N response that year suggest nitrogen limitation. An even stronger case can be made for nitrogen limitations over the comparable interval in 1982. The 26 July enrichment showed a definite nitrogen response, as did the 23 August enrichment. The 10 July enrichment, however, showed a definite response to phosphorus. We therefore conclude that there was a switch from phosphorus to nitrogen limitation in the middle of July in 1982 and probably in 1981 also.

Phosphorus limitation was unequivocal in the June enrichment. The May enrichment, which coincided with spring mixing, also produced a phosphorus response, but the magnitude of the response was much lower at this time, both in absolute and relative terms. We conclude that nutrient limitation was not pronounced in May, and was possibly induced by confinement of the samples in the zone of ideal light for growth. In both years, enrichment during the fall thickening of the mixed layer produced similar results to the enrichments at the time of spring mixing. The absolute and relative magnitudes of the response were small, and the evidence points toward phosphorus limitation more than nitrogen. Under field conditions, it is likely that neither element was immediately limiting at such times, but phosphorus was incipiently limiting and this was brought out by a few days of sample confinement. The October enrichment in 1982 showed a return to pronounced phosphorus control of phytoplankton growth.

The composite picture that emerges from Tables 31 and 32 is summarized in Figure 42. Although no enrichments were done through the ice, the presence of significant inorganic nitrogen under ice and the evidence of phosphorus limitation just before ice formation and just after disappearance of ice strongly support the case for phosphorus limitation under ice. Thus on a calendar year basis the period of phosphorus limitation extended from 1 January until the onset of nitrogen limitation, which appeared rather abruptly in the middle of July. The coincidence of this switch

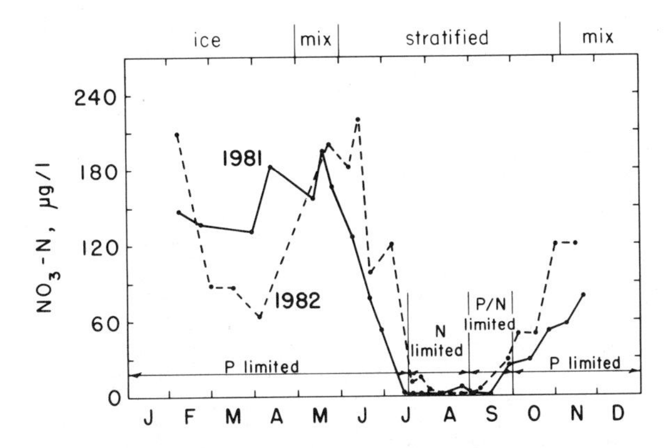

Figure 42. Nutrient limitations as shown by the enrichment experiments superimposed on the nitrate concentrations near the surface (0–5 m).

with the depletion of nitrate is shown in Figure 42. Late summer thickening of the mixed layer relieved nitrogen limitation (mid-September). September can be regarded as a month of transition, during which the response to enrichment was less pronounced but indicative of phosphorus control. Unequivocal return to phosphorus limitation occurred with increased thickening of the mixed layer in October.

The interpretation given here of the enrichment studies is consistent with the inorganic chemistry and with the primary production data. As shown in the previous chapter, P_{max}, an indicator of the nutritional status of phytoplankton, was highest during May or early June and in September. These are periods when the plankton were abundantly supplied with nutrients, and they coincide with the minimal responses to experimental enrichments. P_{max} shows evidence of nutrient stress under ice cover (P depletion) and again in midsummer (N depletion). In the fall there is again evidence of phosphorus deficiency in the P_{max} values.

The enrichment studies showed that phytoplankton biomass in Lake Dillon was controlled part of the year by phosphorus and part of the year by nitrogen. Switching of this type is also thought to occur in Lake Washington (Edmondson 1972). The Lake Dillon switching appears to be predictable in its timing. At the beginning of stratification, the ratio of available nitrogen to available phosphorus was very high. This was determined principally by the high ratio of nitrogen to phosphorus in the water entering the lake. After stratification began, however, the amount of incoming water declined, and much of it entered the middle and deeper layers of the lake, thus not greatly influencing the surface chemistry. Biological processes then gradually changed the ratio of available phosphorus to available nitrogen in the upper water column. From the organic nutrient chemistry, we believe that the mechanism for this change has to do with the relative abilities of phytoplankton to use soluble organic nitrogen and soluble organic phosphorus. As stratification progressed, the relatively large

Table 33. Particulate P Concentrations in the Enrichment Treatments[a]

Date (1982)	Particulate P (μg/l)	
	C, N[b]	P, P + N
24 May	5.0	17.1
14 June	11.9	33.2
10 Jul	6.2	30.2
26 Jul	3.9	29.1
23 Aug	4.5	28.4
7 Sept	3.2	9.8
18 Oct	3.4	22.8

[a] C (control) and N (nitrogen treatment) are together because the SNK test shows them to be consistently indistinguishable for particulate P. The P (phosphorus) and P + N (phosphorus plus nitrogen) treatments are together for the same reason.

[b] C, control; N, N added; P, P added; P + N, P + N added.

amount of phosphorus present in the soluble organic pool gradually disappeared, presumably because it was taken up by the phytoplankton and microbes. This was not true of soluble organic nitrogen, which appeared to be less available to the plankton. Because the phytoplankton and microbes were able to use organic phosphorus more readily than organic nitrogen, the ratio of phosphorus to nitrogen inside the cells changed to such an extent that the phytoplankton became nitrogen limited. Thickening of the mixed layer in September reintroduced abundant supplies of inorganic nitrogen, and thus returned the lake to the phosphorus control that was typical of early stratification.

The effectiveness of the phytoplankton in sequestering and storing phosphorus well beyond their needs is illustrated by the particulate phosphorus data that were taken as part of the enrichment studies. These data are summarized in Table 33. Particulate phosphorus differed significantly among treatments on all dates (one-way ANOVA, $p<0.01$). Phosphorus and phosphorus plus nitrogen treatments were always lumped together by the SNK criteria, and the controls and nitrogen treatments were always lumped together by the same criteria. This was true whether or not the dominant limitation was at a given time principally associated with phosphorus or with nitrogen. Thus the phytoplankton, even when strongly nitrogen limited, took up substantial amounts of the added phosphorus. This lack of linkage between the uptake of phosphorus and the immediate nutritional requirement for phosphorus illustrates the significance of luxury consumption of phosphorus in the Lake Dillon phytoplankton.

10. Horizontal Spatial Variation in the Lake

Up to this point most of the analysis has dealt with samples taken from the index station in the middle of the lake. The question arises how typical the index station is of the lake as a whole. This question is answered here in two stages. First, an analysis is made of the five-station heterogeneity series. As described in Chapters 2 and 3, this series consisted of analyses made on a set of samples taken on 32 different dates at the four main stations and the index station. The top (0–5 m) and bottom of the water column were sampled at each station. Since each one of the four main stations was located at the mouth of one of the four main arms of the lake, this sample series gives information about the degree of variation to be expected over the deep-water section of the lake.

The second stage of the analysis is based on the 14-station heterogeneity series. This series included surface samples (0–5 m) from 14 different sites taken on 10 different dates. Unlike the 5-station heterogeneity series, the 14-station series extended over the entire lake surface without regard to depth, and thus included shallow-water as well as deep-water areas.

The Five-Station Heterogeneity Study

Samples taken at the same depth at multiple stations will always show some degree of variation. This variation can be divided into three components (Platt and Filion 1973, Lewis 1980): (1) error variance, (2) fixed horizontal spatial variation, and

(3) ephemeral horizontal spatial variation. Error variance is that which would be observed in replicates from the same site on a given date. In effect this is analytical error variance, although some handling variance could be included and, if the replicates were taken separately at the same site, some true microscale patchiness might also contribute to the error variance defined in this way. In most cases, and certainly in the case of the variables of concern here, analytical error dominates the variance of replicates from a given site on a given date.

Fixed spatial variation is associated with average differences between stations. For example, if a nearby shoreline or point source of nutrients continuously influences the chemistry of a particular station, this would increase the variance of a set of samples including that station. In contrast, ephemeral spatial variation is due to horizontal variation that is not stable over time. There would be a considerable difference among stations on any given date, but still no difference in the average values for a given variable among the different stations, if ephemeral variation were high and fixed variation were nil. Typically lakes can be expected to show some combination of fixed and ephemeral variation, but ephemeral variation usually dominates except in the most extreme cases of physical or chemical gradients.

For the five-station heterogeneity series, we first consider fixed horizontal spatial variation, i.e., the possibility that the five stations were significantly different when averaged through time with respect to their chemical or biological properties. The analysis is framed as a statistical null hypothesis, which is that all stations have the same average values. The hypothesis is tested by one-way ANOVA.

Table 34 shows the mean value of 11 different variables for samples taken in the 0–5-m layer at each of the five stations. The table also shows the grand mean for each variable over the 32 different dates and five different stations and shows the outcome of the one-way ANOVA testing for significant differences among stations. A cursory examination of the means for the different stations shows that they are remarkably similar to each other. The results of the one-way ANOVA show that in all instances the small amount of variation that is observed from one station to the next is statistically insignificant at $p = 0.05$. We therefore conclude that, from the viewpoint of annual averages, horizontal spatial variation at the surface over the deep part of the lake, including the mouths of the arms of the lake, is trivial. This does not imply that there is no variation between stations on a given day (ephemeral variation), but it does indicate that one of these stations is as good as any other in representing the deep-water portion of the lake over the long run.

The same analysis was done for samples taken at the bottom of the water column at each of the five stations; the results are summarized in Table 35. The list of variables is slightly different than for the surface samples because some analyses that are meaningful at the surface of the water column are not meaningful at the bottom (e.g., secchi depth). The results for the bottom of the water column are tabulated separately because they are subject to a different kind of interpretation. Because the bottom of the water column is found at different depths depending on the station, a certain amount of variation due to depth is blended into the true horizontal variation between stations. Thus while the bottom samples at the index station in the middle of summer were most likely to have been taken at about 40 m, the bottom samples in Dillon Bay were more likely to have been taken at 20 m. Thus there is some variation due to

Table 34. Means for Variables at the Index Station and Four Main Stations on 32 Sampling Dates Spread Over the 2-Year Study Period (0- to 5-m Layer)

Variable	Station						
	Index (A)	Dillon Bay (B)	Snake Arm (C)	Blue Arm (D)	Tenmile Arm (E)	Grand Mean	Significant Differences
Temperature (°C)	9.5	10.2	10.0	10.2	10.3	10.0	None
Secchi (m)	3.0	2.6	2.6	2.7	2.7	2.7	None
Conductance (μmho/cm)	161	162	159	162	162	161	None
Sol. reactive P (μg/l)	1.2	1.0	0.7	0.9	1.0	1.0	None
Total soluble P (μg/l)	3.4	3.2	3.2	2.9	4.0	3.3	None
Particulate P (μg/l)	5.1	5.1	5.2	4.9	5.0	5.1	None
NO_3–N (μg/l)	80	71	77	69	69	73	None
NH_4–N (μg/l)	16	12	14	13	13	14	None
Total soluble N (μg/l)	287	292	275	265	258	275	None
Total particulates (mg/l)	2.9	3.1	3.2	3.0	2.8	3.0	None
Chlorophyll *a* (μg/l)	8.3	8.0	8.2	8.6	8.0	8.2	None

Table 35. Means for Variables at the Index Station and Four Main Stations on 32 Sampling Dates Spread Over the 2-Year Study Period (Bottom of Water Column)

Variable	Station						
	Index (A)	Dillon Bay (B)	Snake Arm (C)	Blue Arm (D)	Tenmile Arm (E)	Grand Mean	Significant Differences
Temperature (°C)	4.3	5.1	4.7	5.0	5.0	4.8	None
Conductance (μmho/cm)	179	167	166	168	171	170	None
Sol. Reactive P (μg/l)	1.0	1.3	1.0	1.0	1.2	1.10	None
Total soluble P (μg/l)	3.0	3.7	3.4	3.5	3.1	3.34	None
Particulate P (μg/l)	3.1	7.8	9.4	6.3	8.6	7.0	Yes
NO_3-N (μg/l)	201	132	141	148	143	153	Yes
NH_4-N (μg/l)	18	25	25	22	25	23.0	None
Total soluble N (μg/l)	387	327	311	423	351	360	None
Total Particulates (mg/l)	1.8	4.9	3.8	2.7	3.4	3.3	Yes

depth alone. No attempt is made here to separate depth variation from the variation due to horizontal patchiness, since the focus of attention is on the nature of the environment near the bottom of the lake.

Table 35 indicates that, despite the additional possibility for variation due to differences in depth between stations, only three variables showed significant differences. These three variables are nitrate nitrogen, total particulates, and particulate phosphorus. In the case of nitrate nitrogen, the variation between stations is explained by the much greater depth of the water column at the index station. Since nitrate nitrogen increased with the depth during stratification, as shown in the seasonal analysis of data for the index station, the bottom samples were taken from a zone richer in nitrogen at the index station than at the other stations. The Student-Newman-Keuls multiple range test segregates the index station from the other four stations at the 5% probability level. The results thus indicate the significance of the rather large pool of nitrate nitrogen very deep in the water column.

For total particulates and particulate phosphorus, the concentrations for the index station are much lower than those for other stations. However, examination of the raw data shows that the higher averages at the four main stations were due entirely to a handful of very high values clustered in the winter of 1981. Since there was very little water movement under ice in the winter, especially in the dry year of 1981, these high values were almost certainly produced by overly close approach of the sampler to the bottom, stirring up sediment and thus inflating the total particulate and particulate phosphorus values. This was not a problem in 1982. Thus for total particulates and for particulate phosphorus, we conclude that the average bottom values for the four main stations in this series are spuriously high.

Even though fixed horizontal variation is not significant either at the surface or at the bottom of the water column with the exception of nitrate in deep water and the spurious cases of total particulate and particulate phosphorus heterogeneity in deep water, ephemeral horizontal spatial heterogeneity may still be significant. A detailed look at the magnitude of ephemeral spatial variance is possible by separation of variance components. The separation will only be done for the surface sample series because of the much greater practical significance of heterogeneity in the growth zone of the lake.

Total variance of surface samples at five stations on a given date (s^2_t) can be broken down as follows:

$$s^2_t = s^2_e + s^2_s$$

where s^2_e is the error variance and s^2_s is the true variance attributable to stations. Since it has been shown that the true variance attributable to stations does not include any significant contribution resulting from fixed (temporally stable) differences between stations, it follows that s^2_s is entirely accounted for by ephemeral spatial variation. Thus if separation of the variance components can be achieved, we will have a quantitative estimate of the magnitude of the ephemeral spatial variation.

The magnitude of error variance for the Dillon samples was obtained as a byproduct of the routine analysis of replicates. For each date and each variable, the total variance among stations was obtained from the raw data and the replicate variance was subtracted from this total, leaving s^2_s. From s^2_s we obtained the standard deviation attributable to ephemeral horizontal spatial variation (s_s). We then calculated the coefficient of variation associated with this ephemeral spatial variation ($s_s/\bar{x} \cdot 100$). The mean values for these statistics are reported in Table 36.

The amount of ephemeral spatial variation, both in absolute terms (s_s) and in relative terms ($s/\bar{x} \cdot 100$) is exceedingly small for all variables. As might be expected, conductance, which is a biologically conservative property, shows the least ephemeral spatial variation, and some of the more biologically sensitive variables show greater ephemeral spatial variation. The overall conclusion is clear: on the average, the degree of horizontal spatial variation over the area covered by the main stations and the index station is very small in relation to the absolute magnitude of the variable that is being measured. A measurement at any one of these stations on a given day tells almost as much as there is to know about important chemical and biological features of the main body of the lake and the mouths of the arms on that day.

When the values of s_s or the coefficients of variation corresponding to s_s are plotted against time, some trends are evident. For all of the variables, there were times of the year when the ephemeral spatial variation was essentially 0, and other times of the year when the ephemeral spatial variation was very definitely above 0, although still small in absolute terms. Table 36 incorporates a summary for each variable of the time spans when the ephemeral spatial variation was maximal in relative terms. Certain patterns are evident from the list of maxima and from the raw data themselves. First, all variables with the exception of total soluble nitrogen showed maxima in relative variation under ice or over a time period that overlaps with ice cover and early spring mixing. The nearly universal tendency toward maxima of relative heterogeneity under ice is explained by the physical condition of the lake at this time. Since the lake surface is not exposed to wind action, any heterogeneity that is induced by outside influences, or by biological phenomena, is much more likely to persist than it would

Table 36. Summary of Variation among the Five Main Stations (0–5 m), after Removal of Error (= Analytical) Variance, and the Months of Peak Variation for Each Variable

Variable	Standard Deviation among Stations[a]	Coefficient of Variation among Stations[b] (%)[a]	Time of Peak Variation
Conductance (μmho/cm)	1.9	1.2	April–May
Sol. reactive P (μg/l)	0.20	8.3	Jan–April
Total soluble P (μg/l)	0.43	11.6	Jan–July
Particulate P (μg/l)	0.26	4.5	Jan–April, August
NO_3–N (μg/l)	4.4	6.4	March–April, July
NH_4–N (μg/l)	1.5	5.9	Jan–April
Total soluble N (μg/l)	33.0	8.7	July
Total particulates (mg/l)	0.12	3.4	April–June
Chlorophyll *a* (μg/l)	0.31	3.9	Jan–April, July–August

[a] Error variance has been removed.
[b] $(s_s/\bar{x}) \cdot 100$.

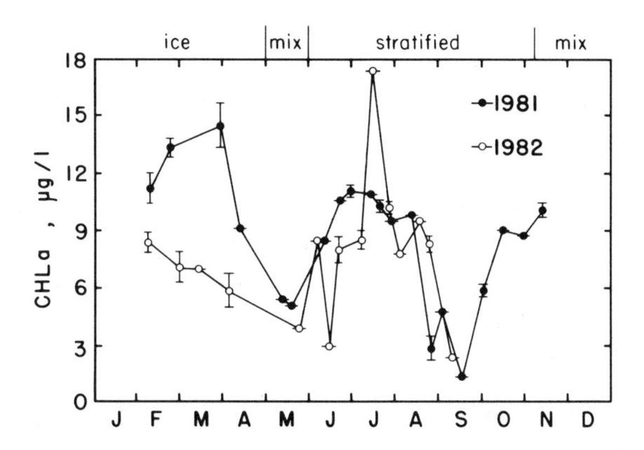

Figure 43. Means on various dates for chlorophyll for the five-station heterogeneity study. For each mean, the vertical bar shows the standard deviation due to true variation between stations (error variance removed).

be under the ice-free condition when the surface water can be homogenized by wind action. Many variables tended to show a period of higher heterogeneity only toward the end of ice cover, and this also corresponds to the physical events that were occurring in the lake. Whereas outside influence on the lake by runoff was absolutely minimal from the formation of ice cover in early January into March, spring weather was accompanied by rapidly increasing external influence in the form of snowmelt, which in its earliest stages passed under the ice cover, thus magnifying the sources of heterogeneity.

Another common tendency was for relative ephemeral spatial variation to increase during late summer. This was true especially of inorganic nitrogen and phosphorus and to some degree also of chlorophyll *a*. At this time of year the phytoplankton were under extreme nutrient stress. Minor physical phenomena could thus induce measurable heterogeneity either in the amounts of soluble inorganic nutrients or in the amounts of phytoplankton chlorophyll. It is important to note that, while these peaks of heterogeneity were undoubtedly real, their absolute magnitude was exceedingly small. By way of illustration, Figure 43 shows the mean chlorophyll values for the five stations over the 2-year period of study and the accompanying estimates of standard deviation due to the spatial component of variance (s_s).

The general conclusion of the five-station heterogeneity study is that the main body of the lake, as encompassed by the index station and the five main stations, can be treated as a unit, and that any one station on a given day is a good indicator of the condition of the entire central region of the lake.

The 14-Station Heterogeneity Study

The 14-station heterogeneity study included not only stations over deep water but also stations near the head of each one of the bays, and a number of stations in the broad Tenmile Arm. The locations of these stations have already been shown (Figure 2). The stations are organized in Table 37 according to location. Table 37 also summarizes the mean values of each one of the variables for each of the 14 stations over the 10 sampling dates covered by the 14-station heterogeneity study. The data for each variable were subjected to analysis by one-way ANOVA according to a rationale identical to that used in the five-station heterogeneity study. Once again, the null hypothesis was that the 14 stations are not significantly different from each other in the mean values for any given variable. In no case could the null hypothesis be rejected at the 5% probability level. We thus conclude that there are no significant differences in the mean values for any of these variables among the 14 stations, despite the coverage of all the different arms and of both shallow-water and deep-water portions of the lake.

Since the ability of a statistical test to discriminate differences is directly related to the amount of data available, extension of the 14-station heterogeneity series to cover 30 or 40 different dates instead of 10 different dates might have shown certain significant differences between stations. The important point, however, is that any such differences were so minor that they could not be detected at all with a 10-date series, which would have been sensitive to consistent differences of any great magnitude between stations.

The raw data and the table of means are suggestive of certain patterns that may be real, although they could not be verified statistically. For example, it would appear from Table 37 that the Tenmile Arm averaged slightly higher in conductance than any of the other arms of the lake or the main body of the lake. The higher conductance of incoming water is largely offset by efficient mixing, however. It is likely that a more extensive data set would be able to show statistically that the Tenmile Arm has a slightly higher conductance on the average than the rest of the lake, but the degree of difference is sufficiently small that biological consequences would be minimal.

The raw data also indicate the tendency of shallow-water stations to show higher values of chlorophyll and lower transparencies during late summer at the time of most aggravated nutrient depletion (July–August). However, since this pattern is sustained for a very short time, it is difficult to demonstrate statistically and must be considered as suggestive rather than definitive. The pattern does make sense biologically insofar as there will be slightly richer supplies of inorganic nutrients where there is more intimate mud–water contact, as in shallow-water areas, and this may introduce a noticeable difference at times when nutrient depletion is extreme.

Another pattern that is highly suggestive but cannot be proven statistically is the tendency of the stations nearest to wastewater treatment plant outfalls to show higher average values for inorganic nutrients, and especially for ammonium. In fact it is remarkable that these stations, because of their proximity to the outfalls, did not show significant differences from the other stations of the lake. Examination of the raw data shows, however, that the sewage effect is, although noticeable, relatively small in magnitude and sporadic in nature. Vigorous wind-driven circulation and variations in the effluents themselves caused the effect to be undetectable more often than not even at the stations closest to the discharges.

The 14-station heterogeneity study argues very strongly that the lake, in terms of general chemistry, nutrient chemistry, and biological variables, can be treated as a unit, and not as a collection of semi-isolated arms and bays that react at grossly different rates to nutrient loading or seasonal events. Although this is counterintuitive to the

Table 37. Results of 14-Station Heterogeneity Study, Showing the Means Over the 14 Stations (0–5 m) for the 10 Sampling Dates Spanning the 2 Years of Study

Station Location	Secchi (m)	Conductance (μmho/cm)	Soluble Reactive P (μg/l)	Total Soluble P (μg/l)	Particulate P (μg/l)	NO_3–N (μg/l)	NH_4–N (μg/l)	Total Soluble N (μg/l)	Particulates (mg/l)	Chlorophyll *a* (μg/l)
Lake Center										
Index station (A)	2.8	164	1.6	3.6	5.2	105	12	227	3.3	7.3
2 km south of index (LS3)	2.5	158	0.6	4.3	5.4	93	11	225	3.5	8.1
Dillon Bay										
Near Dillon (LSI)	2.1	163	0.9	4.0	6.1	76	15	348	5.0	6.7
Mouth of Bay (B)	2.4	167	1.2	3.8	5.2	75	15	251	3.5	7.4
Snake River Arm										
Head of arm (LS2)	2.2	163	0.8	3.2	5.7	86	14	235	4.7	7.3
Mouth of arm (C)	2.4	158	0.6	3.8	5.3	83	12	252	4.3	7.3
Blue River Arm										
Head of arm (LS4)	2.8	164	0.8	3.6	6.8	78	34	256	4.1	8.6
Mouth of arm (D)	2.6	163	0.8	3.2	5.4	89	13	231	3.4	8.0
Tenmile Arm										
Head of arm east (LS9)	2.0	177	1.7	4.6	7.5	53	82	434	3.5	7.1
Head of arm west (LS8)	2.1	201	0.7	3.5	7.5	132	30	361	4.2	6.2
Middle of arm south (LS7)	2.4	175	0.8	3.7	6.0	83	18	215	3.1	7.4
Middle of arm north (LS5)	2.6	172	0.8	3.8	6.2	83	12	256	3.4	7.2
Gibertson Bay (LS6)	2.5	171	0.9	4.1	6.4	78	15	243	3.1	7.6
Mouth of arm (E)	2.6	164	0.7	3.9	5.1	82	11	260	2.9	6.8
Grand Mean	2.4	168	0.9	3.8	6.0	86	20	267	3.7	7.3
Significant differences	None	None	None	None	None	None	None	None	None	None

apparent semi-isolated condition of the arms and the continuous entry of water of different quality into each of the arms, it is obvious that mixing of the lake is sufficiently pronounced under the influence of the wind to homogenize the chemistry and biological variables very efficiently. Perhaps this is not so surprising, in view of the relatively small size and high wind exposure of the lake. Both of these factors act against the development of pronounced or persistent patchiness that would produce significant differences in the arms or between stations.

11. Overview of Limnology and Trophic Status

The Annual Cycle

Figure 44 is a general synopsis of the annual cycle based on the foregoing chapters that have dealt with physical, chemical, and biological variables in Lake Dillon. The cycle is generalized for the 2 years of record. Layering and mixing patterns are the key to understanding chemical and biological changes through the year. Since layering and mixing are principally under the control of seasonal changes of air temperature and solar radiation, the general pattern will be very much the same from one year to the next. This is well illustrated by the similarity of layering and mixing phenomena in 1981 and 1982, which were quite different in amount of runoff.

The predictable duration and timing of ice cover, spring mixing, fall mixing, and summer stratification enforce a certain amount of order in the annual events in the lake. We thus expect the synopsis shown in Figure 44 to have general validity even if the trophic status of the lake should change in the future. Chlorophyll maxima will occur in the month of July, and the months of May and September will show surface chlorophyll minima. The nutritional status of phytoplankton will be most optimal in May and September and least optimal in August. Transparencies will consistently be low in June due to runoff and in July due to phytoplankton, and will be highest in August because of low chlorophyll concentrations resulting from nutrient depletion at that time.

Despite the persistent annual pattern, certain features of the annual cycle are subject to influence by irregular variations, and others are subject to change if the nutrient

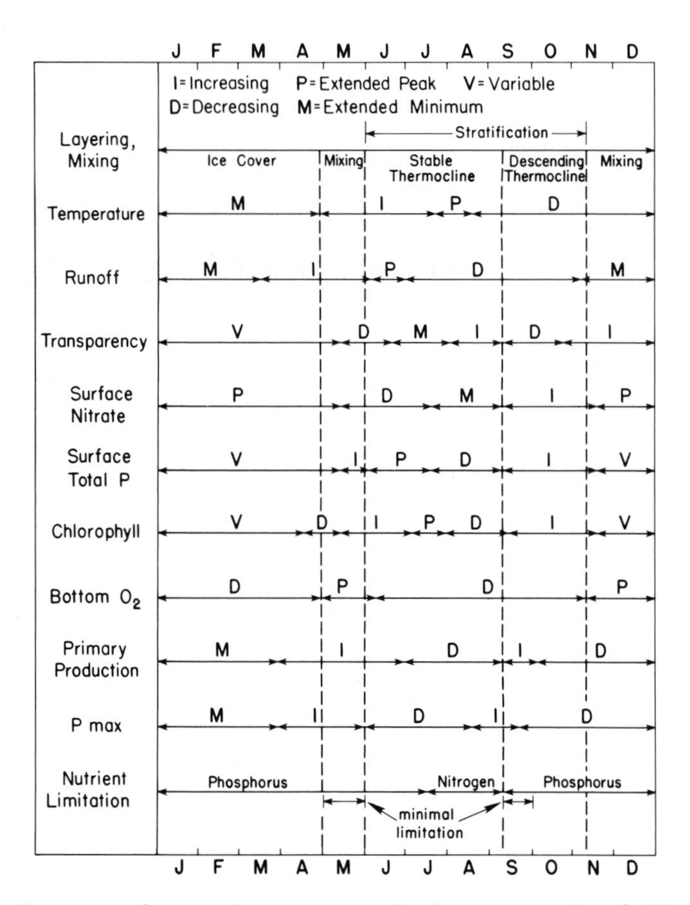

Figure 44. Summary of seasonal events, generalized as a composite of 1981 and 1982.

supply to the lake should change. Runoff is exceedingly variable from one year to the next and it has certain direct influences on the chemistry and biology of the lake. It is runoff, not phytoplankton, that determines the transparency of the lake in June. Higher runoff also raises the phosphorus peak that occurs in June and July, and this will in turn raise the average annual total phosphorus, at least in the upper water column. Counter to intuition, the data for 1981 and 1982 show that higher runoff than average is likely to suppress primary production even though it raises the peak total phosphorus concentration. Inorganic particulates brought in by runoff suppress production by reducing the amount of light that reaches the algae. Higher runoff also maintains a thinner epilimnion and causes injection of the particulates closer to the lake surface. These factors enhance the extinction of light by nonliving particulates, thus discouraging significant production below about 5 m.

Biomass accumulation, as measured by chlorophyll concentrations, is likely to be very similar under high and low runoff conditions, judging from 1981 and 1982. The chlorophyll maximum in a year of low runoff is likely to be situated well below the surface, while the maximum in a year of normal or above-normal runoff will probably include the upper 5 m. On the other hand, the total amount of chlorophyll under a unit of lake surface will be pretty much the same under the two conditions, if the 2 years of record are accurate indicators.

Under the present trophic condition, the phytoplankton of Lake Dillon respond to enrichment with phosphorus plus nitrogen at any time of the year, although the degree of response is minimal in May and September when the upper water column has been freshly enriched with deep water containing nutrients. In 1981 and 1982, there was a critical switch in limiting nutrients from phosphorus to nitrogen in the middle of July, and this is probably a general phenomenon for the lake under its present condition. The phytoplankton use up the soluble reactive phosphorus almost immediately after the onset of stratification. Growth continues nevertheless and is accompanied by steady decline in soluble organic phosphorus. The algae at this time are phosphorus limited, and respond strongly to orthophosphate addition. They have not exhausted the phosphorus supply, however, because they are able to use soluble organic phosphorus, although they obtain it more slowly and less efficiently than they would orthophosphate and thus grow faster if orthophosphate is added. In the middle of July, growth is essentially halted not by phosphorus but rather by inorganic nitrogen, which is present in abundance at the beginning of stratification but is finally used up by the middle of July. The effect of this nitrogen exhaustion is more drastic than the exhaustion of orthophosphate, implying that the phytoplankton are less able to use the soluble organic nitrogen pool than the soluble organic phosphorus pool.

Although phosphorus is presently limiting over a much greater segment of the year than is nitrogen, it is inorganic nitrogen that halts the progress of the summer growth and thus determines the maximum amount of chlorophyll in the epilimnion. Since the summer chlorophyll maximum is a point of major practical concern, it is fair to ask whether eutrophication management strategy should be based on nitrogen rather than phosphorus. Aside from economic considerations, two biological factors should be weighed. First, under the present conditions, the lake is very close to a balance between phosphorus and nitrogen limitation. By the time inorganic nitrogen limitation occurs, phytoplankton have removed not only all the orthophosphate but also virtually all of the soluble organic phosphorus. Large responses to enrichment can be obtained only with a combination of phosphorus and nitrogen. Since the balance is so close, the choice for management is essentially arbitrary and can be based on economic considerations or convenience. A more general version of this argument is given in a recent OECD publication (1982) outlining strategies for eutrophication control.

Another biological consideration is the possible encouragement of nitrogen fixers if the present nitrogen limitation were exacerbated by reduction of nitrogen loading with the same or greater phosphorus loading. As the balance is shifted more and more strongly toward nitrogen limitation, there is an increasing ecological incentive for the growth of algal taxa that have low nitrogen requirements. The algae that have lowest nitrogen requirements are those that can fix nitrogen, specifically the heterocystous blue-greens such as *Anabaena* and *Aphanizomenon.* Any shift toward the dominance of such algae is usually considered undesirable, as these algae also have gas vacuoles that cause them to rise to the surface and form scums. Thus while the July chlorophyll maximum of 1982 gradually sank out of sight as shown in Figure 30, a comparable amount of *Anabaena* biomass might well have come to the surface, where it would have affected the appearance and surface transparency of the lake. This and other considerations will be taken up again later in connection with predictions and modelling.

Trophic Status of the Lake

The trophic status of the lake can be evaluated on the basis of any one or a combination of several trophic indicators. Table 38 lists four indicators for Lake Dillon, and a trophic assignment for each based on traditional boundaries (Welch 1980). The assignment of a given lake to a trophic category is to some degree arbitrary. On the basis of a single trophic indicator, lakes could be arranged in a continuous spectrum that lacks sharp boundaries. Furthermore, the use of several different trophic indicators will typically provide slightly different perspectives on the placement of a particular lake in a trophic sequence.

Phosphorus is a widely used criterion for classifying lakes according to trophic status. Although it has been recognized for decades that phosphorus is of great importance in determining trophic status, an attempt to make a quantitative linkage between phosphorus and trophic status dates back principally to Vollenweider (1965). Vollenweider, following Sawyer (1947), specified that oligotrophic lakes have total

Table 38. Summary of Trophic Indicators for Dillon under Conditions at the Time of the Study[a]

Indicator	Magnitude	Trophic Indication
Total Phosphorus (μg/l)		
Spring surface maximum day	12–16	Lower mesotrophic
Spring surface maximum month	11–13	or
Annual average surface	7–9	upper oligotrophic
July–October average (0–15 m)	7–8	
Transparency (m)		
July secchi minimum day	1.5–1.9	Mesotrophic
July secchi average	1.8–2.3	
July–October average	2.5–3.5	
Chlorophyll *a* (μg/l)		
Summer surface maximum day	10–17	Mesotrophic
Summer surface maximum month	9–12	
July–October average	6–8	
Primary Production		
Annual (g C/m^2/yr)	50–100	Mesotrophic
Maximum daily (mg C/m^2/day)	500–900	
Summer oxygen		
AHOD (mg O_2/m^2/day)	600–700	Lower eutrophic
5 m above bottom minimum (mg/l)	4–5	

[a] Range of magnitudes shows, in round figures, probable span with varying weather in different years.

phosphorus concentrations below 10 μg/liter, that mesotrophic lakes have total phosphorus concentrations between 10 and 20 μg/liter, and that eutrophic lakes have total phosphorus concentrations above 20 μg/liter. No simple rules are available for selection of the appropriate sampling times and the appropriate number of samples for determination of total phosphorus, however. Depending on whether one uses annual or seasonal averages or seasonal peaks, Dillon might currently be classified as lower mesotrophic or upper oligotrophic according to Vollenweider's criteria.

Transparency, chlorophyll, and primary production are also indicators of trophic status. Secchi depths in Dillon are low in both June and July. Table 38 gives the July rather than the June secchi depth as the basis for judging trophic status because the low secchi depths in June are due to inorganic particulates rather than chlorophyll. Secchi depths in July are low enough to bring the lake into the mesotrophic category by traditional criteria (Likens 1975, Welch 1980). Similarly, chlorophyll *a* maxima and growing season averages indicate mesotrophic status. In fact the chlorophyll *a* values are higher than in most lakes that have comparable total phosphorus values (e.g., Kalff and Knoechel 1975). The reasons for this will be explored in connection with the development of the model. Primary production, considering the short growing season, is also in the mesotrophic range.

Areal hypolimnetic oxygen deficit is a potential trophic indicator but is less often used than the other indicators of Table 38. Mortimer (1941) set boundaries of 250 and 550 mg O_2/m^2/day dividing the three trophic categories. Thus according to this indicator, Lake Dillon is just entering the eutrophic category.

In summary, the best characterization of Lake Dillon at the present time seems to be mesotrophic. The lower mesotrophic boundary is an important one, as changes in the lake are most visible to the casual observer over the mesotrophic range.

12. Chemistry of Nutrient Sources as They Enter the Lake

Pathways followed by water and nutrients entering Lake Dillon can be divided for present purposes into six categories: (1) major rivers, (2) streams not joining a major river prior to reaching the lake, (3) sewage effluents, (4) miscellaneous surface drainage not accounted for in other categories, (5) precipitation, and (6) groundwater. All of these categories except miscellaneous surface drainage and groundwater, which made minor contributions, were studied chemically at or very near their point of entry to the lake. In the first category are the Snake River, the Blue River, and Tenmile Creek. These will take up most of the attention of this chapter, since they account for the bulk of surface transport to the lake. In the second category are Soda Creek and Miner's Creek, both of which were sampled above the point sources that enter them near their mouths. In the third category are the effluents of the Frisco WWTP (which enters Miner's Creek near the lake) and Snake River WWTP (which enters Soda Creek near the lake). Both of these effluents were sampled prior to entering streams. Breckenridge effluent was part of the Blue River sample, since the effluent enters above the river mouth, and Copper Mountain effluent was part of the Tenmile Creek sample for the same reason. Discharge data for rivers were obtained from USGS data records. For the effluents, discharge data were obtained from plant operators, and for Miner's Creek and Soda Creek the data were based on our own measurements. All chemistry data were from our own analyses.

The chemistry of the nutrient sources that were sampled routinely will be considered here. Contributions of groundwater and miscellaneous surface drainage will be estimated in Chapter 13, which gives total transport, and the breakdown of surface sources above their points of entry to the lake will be made in Chapters 14 and 15, which deal with land use in relation to nutrient yield.

24-Hour Variation in River Chemistry

The mouths of the major rivers were sampled on 54 different dates over the 2-year period. Since the samples were taken mostly during the daylight hours, a question arises as to whether there is any regular periodicity in stream chemistry that might cause this type of sampling to introduce bias. Although no major periodicity was expected *a priori,* two 24-hr studies were undertaken in order to show whether periodicities were of any concern. A secondary purpose of these studies was to show the degree of short-term irregular variation, which would contribute to random noise in the data but would not bias the results of a large data set.

On 14 January 1981, samples were taken at each of the three river mouths at 3-hr intervals beginning at 1600 and ending at the same time the next day. These samples were analyzed for total particulates, ammonium, nitrite plus nitrate, and soluble reactive phosphorus. The study was repeated in November 1981 with a slightly different sampling interval (4 hr). In the November sampling, analyses were made for total soluble phosphorus, total soluble nitrogen, particulate phosphorus, and total particulates. In November 1981 the samples of the Blue River were taken above the Breckenridge Wastewater Treatment Plant, and in January 1981 they were taken below it. In November 1981 the Tenmile Creek samples were taken above Frisco and in January they were taken below Frisco.

The results of the two 24-hr studies are analyzed here in two stages. First, the data series for each variable at each river mouth on each date is tested for the existence of significant additional variance above and beyond that which would be expected from analytical variance alone. The presence of significant additional variance indicates either irregular or patterned variation in the stream, or some combination of these. If additional variation is present, the distinction between irregular and patterned variation will be made on the basis of subjective evaluation of pattern in the data, since the time series is too short to allow statistical time series analyses.

Table 39 gives the mean for the variables over the 24-hr sampling interval at each of the three sites on each of the two dates. In addition, the standard deviation of each series is shown. The ratio of the square of the standard deviation to the error (= analytical) variance is the F statistic, which is also given. The F statistics for each set of analyses were compared with critical values for F at the 0.05 probability level (Rohlf and Sokal 1969). On this basis, F values indicating sample variance significantly in excess of the analytical error variance were identified and have been marked with asterisks in the table. In instances where the F statistic gives evidence of significant true variation in stream chemistry over the 24-hr cycle, the variation may have been either patterned or irregular.

Significant variation in stream chemistry above the error variance did not occur for any of the analyses at any of the three rivers in the November sample series, so we conclude without further analysis that on this date a sample taken at any time of day would have represented without significant bias the chemistry of the river from which

Table 39. Means and Standard Deviations for the Two 24-hr Sampling Series, and the F Statistic for Each Series

	Snake River			Blue River			Tenmile Creek		
	Mean	SD	F	Mean	SD	F	Mean	SD	F
January 1981									
Particulates (mg/l)	1.46	1.40	6.32[a]	0.98	0.44	0.62	1.11	1.84	10.7[a]
NH_4-N (μg/l)	10.3	6.4	1.30	434	223	1570.0[a]	495	81	209.0[a]
NO_2 + NO_3-N (μg/l)	124	8.0	1.83	577	417	4995.0[a]	1006	85	207.0[a]
Sol. reactive P (μg/l)	0.33	0.61	2.32	1.45	1.51	14.0[a]	0.77	0.68	2.89[a]
Particulate P (μg/l)	1.71	1.13	3.79[a]	3.6	0.53	0.83	2.63	1.65	8.09[a]
November 1981									
Particulates (mg/l)	1.62	0.61	1.19	1.84	0.53	0.90	0.79	0.65	1.34
Total soluble N (μg/l)	263	36.0	0.21	118.0	39.0	0.25	117.0	20.0	0.06
Particulate P (μg/l)	1.1	0.70	1.45	2.9	0.2	0.11	1.8	0.6	1.07

[a] Sample variance exceeds error variance significantly ($p<0.05$).

it was taken. In January, however, there was significant variation in stream chemistry over the 24-hr cycle for about two thirds of the combinations of variables and collection sites. Tenmile Creek showed significant variation for all variables, the Blue River showed significant variation for three of the four variables, and the Snake River showed significant variation for only one variable.

Figure 45 depicts the stream chemistry values for each combination of sites and variables in the January series where there was evidence of a significant variation in stream chemistry over the 24-hr sampling period. The data do not suggest any kind of repeated pattern. For example, there was a definite minimum of ammonium on the Blue River at 4 P.M. one day, but the next day the ammonium concentration was high

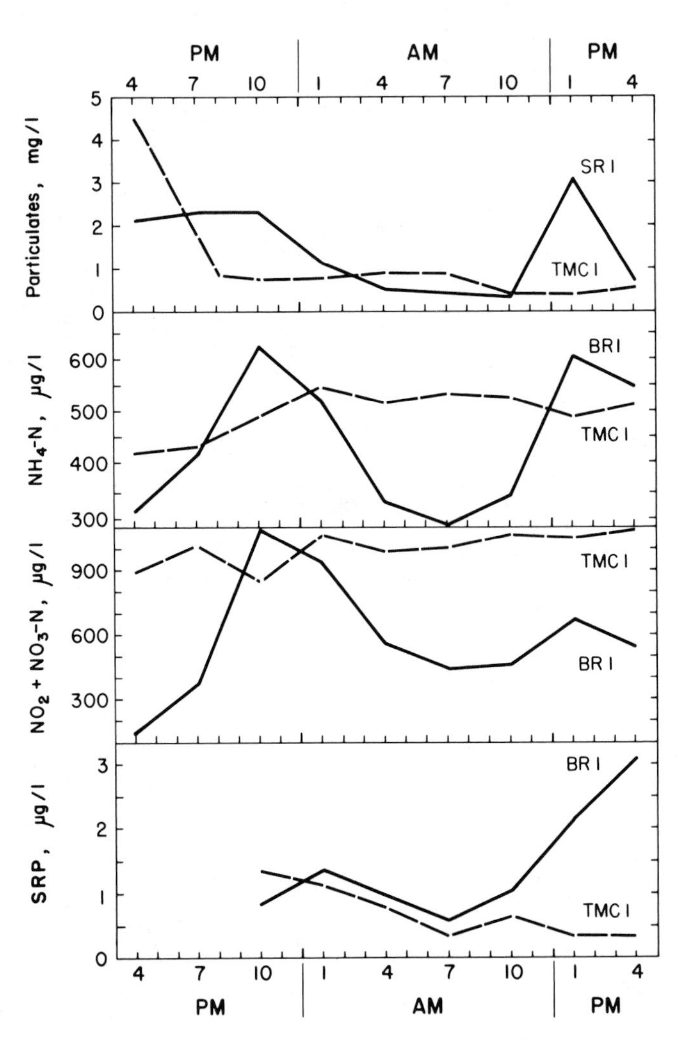

Figure 45. Data for variables that showed statistical evidence of significant diel variation during the study of January 1981 (SR1, Snake River; TMC1, Tenmile Creek, BR1, Blue River).

at 4 P.M. Although the data set is not large enough to allow rigorous statistical analysis for pattern, we conclude with reasonable certainty that the added variance in each case is due to randomly timed irregularities and not to predictable patterns in stream chemistry. This being the case, samples taken at any time of the day or night will represent the stream without bias, although a larger sample series is required to yield a given confidence interval for transport when added variance is present.

Chemistry of the Three Rivers as They Enter the Lake

Table 40 summarizes the average concentrations of dissolved and suspended materials in the three rivers as they enter the lake. Averages can be obtained by unweighted averaging, time-weighted averaging, or discharge-weighted averaging:

Unweighted averaging $$\overline{C} = \sum_{i=1}^{n} C_i / n$$

Time-weighted averaging $$\overline{C} = \sum_{i=1}^{n} C_i t_i / \sum_{i=1}^{n} t_i$$

Discharge-weighted averaging $$\overline{C} = \sum_{i=1}^{n} C_i D_i / \sum_{i=1}^{n} D_i$$

where $\overline{C}$ is mean concentration, C_i is the concentration at the ith time interval, t_i is the length of the interval, and D_i is the discharge over the interval.

In the present case, unweighted averaging is not very defensible because the samples were taken more frequently at times of high discharge when concentrations of dissolved and suspended constituents were likely to be changing most rapidly. A time-weighted average would give the best idea of the chemical composition of the rivers without regard to seasonal changes in discharge, but the discharge-weighted average is the best representation of the chemistry of the entire annual flow accumulated in proportion to discharge, as it actually is in the lake. Thus Table 40 shows the discharge-weighted averages. Time-weighted averages might be useful for certain purposes, but are not tabulated because they are in most instances very close to the discharge-weighted averages. There are a few exceptions to this. For example, total soluble nitrogen is as much as 25–50% higher when expressed as a time-weighted average than it is when expressed as a discharge-weighted average due to some high winter concentrations. Total particulates and all particulate constituents are consistently higher when expressed as discharge-weighted averages than when expressed as time-weighted averages. This is explained by a general tendency of particulate loads of running waters to increase as a power function of discharge (Leopold et al. 1964, Bormann et al. 1969). In contrast, some dissolved constituents show a tendency toward higher means for time-weighted averages than for discharge-weighted averages because of dilution effects. This is especially true of point sources, but also to some degree of nonpoint sources and of certain constituents of natural soil-water systems (Lewis and Grant 1979).

Table 40. Discharge-Weighted Means and Standard Errors of Means for Water Chemistry Variables in the Three Rivers at Their Point of Entry to the Lake

	1981						1982					
	Snake River		Blue River		Tenmile Creek		Snake River		Blue River		Tenmile Creek	
	Mean	SE	Mean	SE	Mean	SE	Mean	SE	Mean	SE	Mean	SE
NO_2–N (μg/l)	0.9	0.12	10.0	1.92	3.5	2.4	1.4	0.58	3.2	0.79	11.0	1.5
NO_3–N (μg/l)	95.0	6.7	227.0	43.7	212.0	30.0	97.0	10.6	131.0	17.0	188.0	25.1
NH_4–N (μg/l)	11.0	1.9	187.0	36.0	26.0	13.0	12.0	1.6	107.0	32.5	94.0	13.9
Total soluble N (μg/l)	183.0	32.3	551.0	104.0	357.0	78.0	234.0	30.6	382.0	74.7	476.0	69.1
Particulate N (μg/l)	33.0	6.8	28.0	5.4	66.0	16.0	107.0	34.9	73.0	17.9	89.0	17.9
Sol. reactive **P** (μg/l)	1.2	0.15	14.0	2.55	1.1	0.14	2.7	0.50	4.3	0.68	2.1	0.37
Total soluble **P** (μg/l)	2.0	0.21	14.0	2.89	3.6	0.44	6.3	1.2	6.7	0.69	5.5	0.73
Particulate **P** (μg/l)	4.2	0.53	5.5	0.64	7.1	1.7	6.9	1.0	5.6	0.54	5.9	0.63
Total particulates (mg/l)	4.0	0.79	2.6	0.48	10.0	3.0	6.1	0.81	3.8	0.37	4.5	0.58
Particulate C (μg/l)	252.0	38.7	194.0	21.4	691.0	195.0	451.0	91.0	184.0	24.3	304.0	44.7
Alkalinity (mg/l)	17.0	1.1	44.0	0.94	35.0	2.18	23.0	1.9	40.0	1.3	39.0	1.9
pH	7.0	0.05	7.6	0.09	7.3	0.07	7.4	0.03	7.8	0.05	7.8	0.05
Conductance (μmho/cm)	101.0	2.55	168.0	3.14	190.0	25.3	84.0	4.2	134.0	3.5	236.0	18.0
Discharge (l/sec)	1145.0	*a*	1585.0	*a*	1839.0	*a*	2334.0	*a*	3113.0	*a*	3299.0	*a*
Discharge (cfs)	44.0	*a*	56.0	*a*	65.0	*a*	92.0	*a*	110.0	*a*	117.0	*a*

[a] Continuous record from USGS.

A detailed consideration will be presented in Chapter 15 of the relative contributions of point and nonpoint sources to the total transport of the three rivers. For the time being, however, it suffices to note that the distribution of land uses and point sources differs among the three rivers. The Snake River is not influenced by any major point source, although it does reflect the influence of nonpoint source nutrient loading from ski slopes and residential development. The Blue River is influenced by a variety of nonpoint sources and also by a major point source, the Breckenridge Wastewater Treatment Plant. Tenmile Creek is influenced by mining operations at Climax Molybdenum Mine in the headwaters, plus a variety of nonpoint sources between the headwaters and the mouth. In addition, Tenmile Creek is influenced by the Copper Mountain Wastewater Treatment Plant, which is situated about 10 km upstream from the sampling point at the mouth (Figure 3).

Table 40 shows, as might be expected from the foregoing synopsis of the three watersheds, that the soluble inorganic nitrogen and total soluble nitrogen concentrations for the Snake River in 1981 and 1982 were considerably lower than those of the other two rivers. For all three rivers, half to three quarters of the soluble nitrogen was inorganic. Particulate nitrogen was generally present in concentrations much lower than soluble nitrogen. Only in the Snake River in 1982 did the particulate nitrogen approach one third of the total nitrogen.

Contrasts among the three watersheds were not so extreme for phosphorus compounds as they were for nitrogen compounds. Soluble reactive phosphorus was considerably higher in the Blue River than in either of the other two rivers. This was largely due to the proximity of the treatment plant to the collection point. The 1981 concentration was exceptionally high, and requires special explanation. A scatterplot of the concentration of SRP in water of the Blue River shows that the very high average for 1981 was caused by exceptionally high values of SRP for the interval September through November of 1981 (Figure 46). These high concentrations were not repeated in the following year, nor did they appear at the next station upstream on the Blue River between September and November of 1981. It is thus clear that the treatment plant was responsible for the exceptionally high average of SRP in 1981. A treatment plant malfunction was confirmed by the plant operators. Figure 46 is thus an illustration of the effectiveness of present tertiary treatment practices.

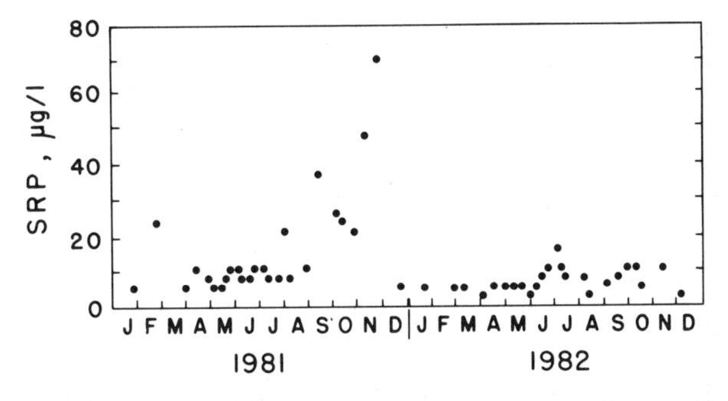

Figure 46. Soluble reactive P in the Blue River at its point of entry into Lake Dillon, showing the effect of malfunctions in tertiary treatment at the Breckenridge WWTP, September–November 1981.

Discounting the high values for Blue River 1981, the concentrations of both soluble reactive and total soluble phosphorus tended to be higher in 1982, a wet year, than in 1981, a dry year. A similar trend has been shown for an undisturbed watershed near the Continental Divide (Lewis and Grant 1979). Higher streamwater concentrations in 1982 are consistent with the higher total soluble phosphorus concentrations in the lake in 1982. The higher concentration of soluble phosphorus in wet years is postulated by Lewis and Grant (1979) to be caused by a critical shift in the balance between biological phosphorus demand, which sequesters P, and the influence of physical forces, which remove soluble P from the zone of demand.

The particulate phosphorus and total particulate concentrations of Tenmile Creek were higher in 1981 than in 1982, contrary to what might have been expected. This was due mostly to high particulate concentrations observed at peak runoff in 1981 but not in 1982. This phenomenon was observed not just at the mouth of Tenmile Creek, but also at other points on the Creek. Its explanation will be considered along with the segment-by-segment analysis of Tenmile Creek to be given in Chapter 15.

Conductance, representing total soluble ionic materials, shows the expected dilution effect between years of high discharge and years of low discharge for the Snake River and Blue River, but not for Tenmile Creek. In addition, the data show clearly

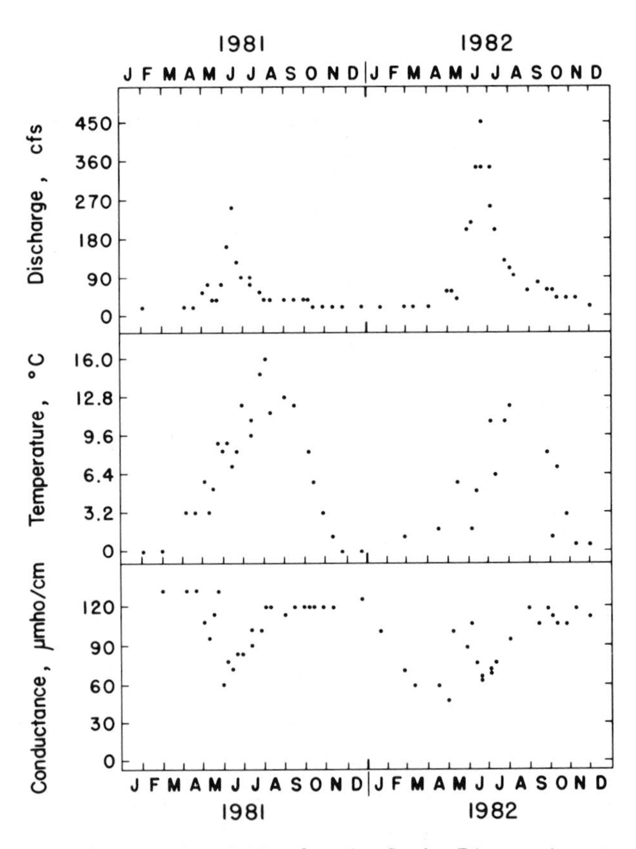

Figure 47. Values of selected variables for the Snake River as it enters Lake Dillon.

that the conductance of Tenmile Creek averaged considerably higher than that of either of the other two rivers. Major sources of ionic dissolved solids associated with mining are located near the head of Tenmile Creek. These raise the average conductance for Tenmile Creek and produce irregularities that are not part of the natural seasonal cycle.

Most of the dissolved and suspended materials showed some kind of seasonal pattern in all three rivers. The natural seasonal pattern is more difficult to pick out for the Blue River and Tenmile Creek than it is for the Snake River because of the influence of point sources. We therefore use the Snake River to illustrate the background seasonal patterns, and then describe how the other two rivers differ from the Snake River.

Figures 47-49 show the seasonal patterns of selected variables over the two years of study for the Snake River. Some variables have been omitted because their indications are redundant or because they are not of special interest here.

Discharge is a key variable, not only because it embodies the seasonal changes that affect concentration but also because it is multiplied by concentrations in the process of obtaining discharge-weighted concentration averages. The pattern shown by discharge in Figure 47 for the Snake River is reflected very closely in the discharge parterns of the other two rivers. The discharge patterns were strongly dominated by

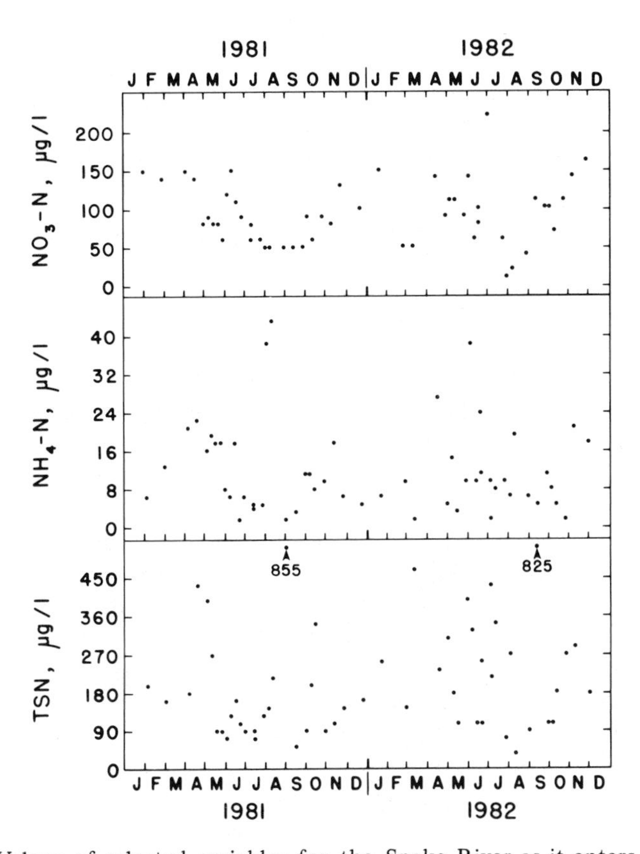

Figure 48. Values of selected variables for the Snake River as it enters Lake Dillon.

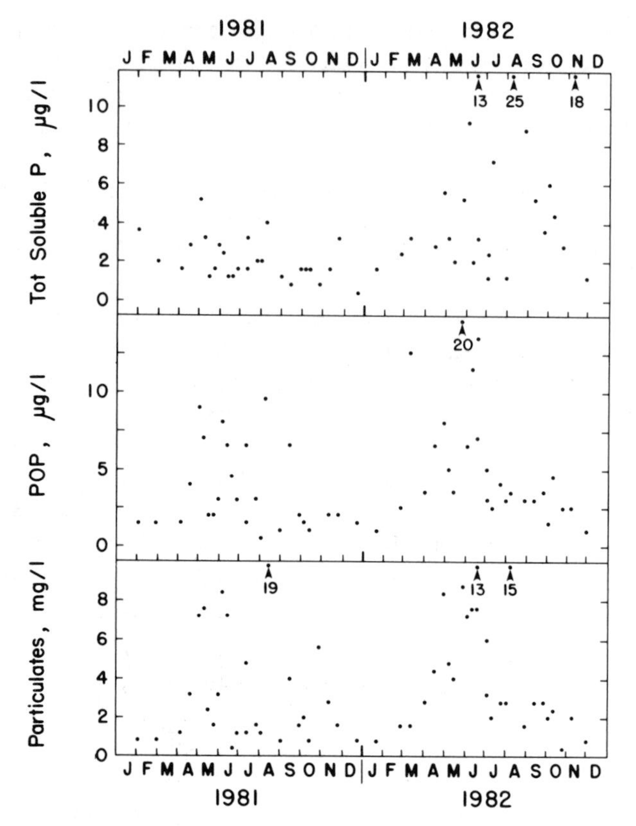

Figure 49. Values of selected variables for the Snake River as it enters Lake Dillon.

seasonal runoff associated with snowmelt. In 1981, when total annual discharge was low, discharge peaked in the first week of June. The peak was broader in 1982 and the highest values did not occur until the end of June. In both years there was a small early peak in May, probably due to early melt in the valley bottoms, and a shoulder or low peak in August or September. The amount of variation around the seasonal pattern was small, which provides an ideal situation for the estimation of weighted mean concentrations and transport.

Stream temperatures varied between 0 and 16°C. The highest temperatures occurred during the last half of July. Maximum temperatures were lower in 1982 than in 1981 by 2–4°C. The seasonal change in temperatures was relatively smooth, although some irregularities were caused by variations in time of day when temperature was taken. Variation of temperature with season and between years affected the depth of penetration of water entering the lake (Figure 19).

Conductance showed the dilution effects that one would expect to be associated with changes in discharge for materials that are soluble and little affected by biological activity. For the Snake River, the conductance closely reflected the early small peak in discharge (May), the subsequent brief decline in discharge, and the major peak of discharge caused by spring runoff. This sequence of events was visible both years for the

Snake River and was also characteristic of the other two rivers. The winter conductances of Tenmile Creek were especially high, however, because of the presence of especially potent sources of soluble materials in the upper part of this watershed at Climax Molybdenum tailings ponds.

Nitrate concentrations of the Snake River were high at the time of runoff and then decreased steadily until August, after which they increased steadily until December or January. Midwinter concentrations were typically high. The low nitrate concentrations in the late summer months were almost certainly due to high summer biological demand for inorganic nitrogen, which became all the more effective in reducing nitrate levels as the amount of water movement declined at the end of summer. With the onset of cold weather, and reduced biological demand, nitrate concentrations again rose. The winter nitrate levels of the Snake were somewhat higher than would be expected for unaltered watersheds in this region (cf. Lewis and Grant 1979, 1980a, 1980b). They probably reflect the influence of dispersed nutrient sources, especially from residential areas, which have an especially noticeable influence on winter concentrations because of the very low discharges in winter and the low biological demand for nitrogen at this time. The seasonal pattern of nitrate concentrations observed in the Snake River is, however, essentially what would be observed in an unaltered Rocky Mountain watershed (Lewis and Grant 1980b). The same pattern is visible in the data for Blue River and Tenmile Creek, but with considerable scatter due to the influence of point sources. The late summer minimum and winter maximum are evident, however.

Tenmile Creek was unusual in that nitrate concentrations were higher at all times of the year than in other rivers. In 1981 nitrate nitrogen approached 1,000 μg/liter in winter (Figure 50). The high nitrate concentrations are explained by the large amounts of nitrate originating at Climax Molybdenum mines at the head of Tenmile Creek. This special nitrogen source will be considered further in the segment-by-segment analysis of Tenmile Creek.

For ammonium, there was also a seasonal pattern of concentrations in 1981 and 1982, but it was not so clear as the pattern for nitrate. In part this was due to the much lower concentrations of ammonium than of nitrate, especially in the Snake River. The Snake River data do show indications of an early spring maximum in concentration before discharge had begun to increase drastically, a late summer minimum

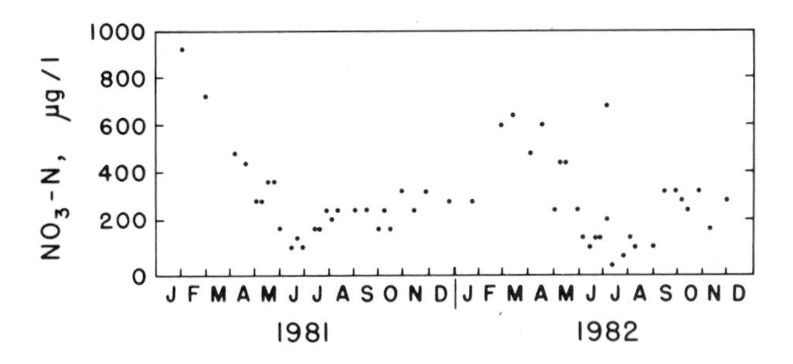

Figure 50. Nitrate values for Tenmile Creek at its point of entry to Lake Dillon, illustrating the very high nitrate levels at certain times of the year.

corresponding to the postulated peak of biological demand for inorganic nitrogen, and a fall increase after the stream water had begun to cool. Two high values in early August of 1981 for the Snake River departed from the seasonal pattern and were not repeated in 1982. These high values were confirmed, however, by other samples upstream and were almost certainly caused by some temporary or sporadic man-made source.

Patterns in the ammonium data for the Blue River and for Tenmile Creek are scarcely discernible because of the large amount of variability introduced by point sources. Ammonium concentrations were affected simultaneously not only by seasonal factors but also by variations in release from point sources and variations in the conversion rate from ammonium to nitrate.

In all three rivers, total soluble nitrogen was accounted for approximately half by organic nitrogen and half by ammonium plus nitrate. Winter values of total soluble N tended to be higher than summer values. Seasonal pattern was weak, however, especially for the Blue River and Tenmile Creek.

Total soluble phosphorus concentrations showed no strong seasonal pattern in the Snake River, nor did SRP concentrations. The average concentrations were often high at times of high discharge, however. The implication is that dilution effects associated with major seasonal increases in discharge were effectively cancelled by increased transport at such times (cf. Lewis and Grant 1979). There were no clear seasonal patterns in concentration for the Blue River or Tenmile Creek, where the point sources introduced considerable additional random variation. For the Blue River, the increase of soluble reactive and total soluble phosphorus in September, October, and November of 1981 caused by treatment plant malfunctions has already been mentioned. A similar but less exaggerated phenomenon was observed in Tenmile Creek in the first half of 1981, but was not repeated in 1982 and was of obscure origin.

Total particulates, particulate phosphorus, particulate nitrogen, and particulate carbon show patterns similar to each other. In the Snake River there was a trend toward high concentrations around the time of maximum discharge. Superimposed on this was an occasional exceptionally high value associated with some kind of short-lived earth disturbance in the watershed, as shown for the Snake River in late August of 1981 and in the summer of 1982. Such short-term spikes obviously can be important in total annual transport of particulate materials.

Chemistry of Miner's Creek and Soda Creek

Because of the small proportion of total land area drained by Miner's Creek and Soda Creek, the potential contribution of these two watersheds to total loading of the lake with nutrients is small. Their chemistry is considered here mainly for the sake of completeness, since these two watersheds are separate from the three major river drainages. Table 41 gives the discharge-weighted averages for the chemistry of the two streams and the standard error for each variable. As in the case of the three rivers, the time-weighted means were calculated but are not shown in the table. For Miner's Creek and Soda Creek, the time-weighted means are even closer to the discharge-weighted means than for the three major rivers.

Table 41. Discharge-Weighted Means for the Two Small Streams Entering the Lake (Sampling Stations above WWTP Effluents)

	1981				1982			
	Miner's Creek		Soda Creek		Miner's Creek		Soda Creek	
	Mean	SE	Mean	SE	Mean	SE	Mean	SE
NO_2–N (μg/l)	0.8	0.11	1.2	0.32	1.4	0.20	2.3	0.23
NO_3–N (μg/l)	32.2	4.3	12.7	4.5	20.2	4.1	63.0	14.0
NH_4–N (μg/l)	14.9	3.0	12.2	2.8	10.0	1.9	11.5	1.4
Total soluble N (μg/l)	141.0	34.0	90.0	21.0	102.0	26.0	202.0	41.0
Particulate N (μg/l)	34.0	13.0	56.0	17.0	83.0	16.0	100.0	29.0
Sol. reactive P (μg/l)	1.9	0.23	8.7	1.2	2.9	0.47	13.4	2.7
Total soluble P (μg/l)	3.2	0.34	18.3	3.3	5.1	0.60	20.5	3.0
Particulate P (μg/l)	6.4	1.2	11.0	1.5	5.8	0.76	21.6	4.0
Total particulates (mg/l)	6.9	1.9	6.1	1.2	7.7	1.7	16.3	3.1
Particulate carbon (mg/l)	320.0	70.0	351.0	63.2	531.0	116.0	501.0	103.0
Alkalinity (mg/l)	28.9	1.7	59.0	3.2	26.0	1.2	62.3	2.4
pH	7.4	0.08	8.0	0.15	7.5	0.05	8.0	0.08
Conductance (μmho/cm)	74.0	3.4	168.0	9.8	58.0	2.6	142.0	8.5
Discharge (l/sec)	88.0	10.0	12.0	3.1	246.0	47.0	54.0	15.0
Discharge (cfs)	3.1	0.37	0.42	0.11	8.7	1.7	1.9	0.52

The small size of contributions of these two creeks to total loading of the lake is evident from the discharge figures in Table 41. The chemical data are unexceptional. Soluble nitrogen concentrations were in the range of those observed in the Snake River, and considerably below those of the Blue River and Tenmile Creek as reported in Table 40. This is not unexpected, since Miner's Creek and Soda Creek at the sampling points were not influenced by point sources of nutrients, but were influenced by some dispersed human influences analogous to those characteristic of the Snake River drainage.

For Miner's Creek, the phosphorus concentrations, including both soluble and particulate fractions, were relatively low, and were comparable to the values observed in the Snake River. For Soda Creek, however, both particulate and total soluble phosphorus concentrations were higher than for any of the three major rivers, and the discharge was especially low in relation to drainage area. Conductance was also notably higher for Soda Creek than for Miner's Creek, probably because of geologic sources for which Soda Creek was named.

Frisco and Snake River Treatment Plant Effluents

Table 42 summarizes the chemistry of the effluents from the Frisco and Snake River wastewater treatment plants, both of which use tertiary treatment for phosphorus. Total soluble nitrogen concentrations were 50 to 100 times higher than the stream concentrations and were similar between plants and between years. The phosphorus values present quite a different picture. First, the effluents of the two plants were rather different in the partitioning of total phosphorus. In the Snake River Wastewater Treatment Plant effluent, the largest amount of phosphorus was in the particulate fraction, while in the Frisco effluent the largest fraction was in the soluble fraction. For the Snake River Plant, the phosphorus concentrations were about 10 times above those of the receiving stream. A few exceptionally high concentrations were checked against plant records. High values typically coincided with construction or equipment problems. The average total P concentrations for the Frisco Plant were considerably higher than those for the Snake River Plant, and there was a major difference between 1981 and 1982 for the Frisco Plant. In 1982 the discharge-weighted averages were very high at the Frisco Plant. The high averages were caused by discontinuous but repeated occurrence of very high concentrations (1000 to 3500 μg P/liter). Checks of the dates of high values with plant operators showed that, for the most part, the dates of high P values coincided with known equipment problems.

Atmospheric Deposition at the Main Station

The atmospheric deposition data consist of the continuous record from the collector situated near the Snake River Wastewater Treatment Plant and shorter records from two other locations. We will consider the more extensive record first and then compare it with the other two records.

Table 42. Discharge-Weighted Means for the Two Effluents Discharging Near the Mouths of Miner and Soda Creeks

	1981				1982			
	Frisco WWTP Effluent		Snake WWTP Effluent		Frisco WWTP Effluent		Snake WWTP Effluent	
	Mean	SE	Mean	SE	Mean	SE	Mean	SE
NO_2-N (μg/l)	135.0	15.8	116.0	17.0	113.0	25.0	268.0	50.0
NO_3-N (μg/l)	1579.0	407.0	3527.0	786.0	1228.0	281.0	2251.0	910.0
NH_4-N (μg/l)	2787.0	670.0	1825.0	620.0	2868.0	543.0	1957.0	453.0
Total soluble N (μg/l)	8911.0	1168.0	6798.0	774.0	7096.0	867.0	8398.0	1627.0
Particulate N (μg/l)	167.0	23.0	496.0	160.0	—	—	—	—
Sol. reactive P (μg/l)	53.6	18.0	4.5	2.4	496.0	169.0	24.4	12.0
Total soluble P (μg/l)	76.7	19.0	22.0	3.4	556.0	177.0	38.6	13.0
Particulate P (μg/l)	17.6	2.2	45.9	11.2	57.0	16.0	120.0	32.0
Total particulates (mg/l)	1.3	0.28	6.0	2.4	5.1	1.2	9.3	2.2
Particulate carbon (mg/l)	741.0	163.0	4710.0	2384.0	—	—	—	—
Alkalinity (mg/l)	42.2	2.9	50.5	7.9	65.8	7.7	68.5	6.9
pH	6.8	0.08	6.5	0.16	7.2	0.10	7.0	0.09
Conductance (μmho/cm)	417.0	29.0	450.0	18.0	449.0	21.0	429.0	20.0
Discharge (l/sec)	12.8	0.89	13.4	0.92	14.5	0.50	14.6	1.2
Discharge (cfs)	0.5	0.02	0.5	0.04	0.5	0.02	0.5	0.04

All of the data to be dealt with here refer to bulk precipitation, which is the total of all wet and dry deposition on a collecting surface. The collecting surface accumulates dry deposition when there is no rain or snow. Wet precipitation augments deposition, especially for certain substances that are scrubbed out of the atmosphere effectively by water droplets (e.g., nitrate). Because of the continuous deposition of materials on a collector surface during dry weather and the variable concentrations during a wet precipitation event, concentration data are not very meaningful for bulk precipitation. For this reason, we express all the results in terms of deposition rates, i.e., the amount of material delivered to a unit surface over a specified time.

Table 43 summarizes the means and standard errors for deposition at the main precipitation collector. The means are derived from 61 separate collections spread over the 2-year study period. Since the collection intervals were not all of identical length, the means are time-weighted averages.

The hydrogen-ion deposition was lower (4X) than that reported for the Como Creek watershed at similar elevation near Ward, Colorado (Grant and Lewis 1982), but precipitation pH was frequently below the bicarbonate-CO_2 equilibrium, indicating the presence of strong mineral acids. Because of the dominant influence of streams on the chemistry of a lake such as Dillon, however, direct hydrogen ion deposition on the lake surface is of minimal significance, even if hydrogen ion deposition is well above background. Nitrate was deposited in substantial amounts at Lake Dillon. As at Como Creek (Grant and Lewis 1982), the precipitation at Lake Dillon appears to be augmented beyond background in nitric acid as a result of the presence of substantial amounts of NO_X in the atmosphere from fossil fuel combustion. The ammonium deposition was also high but secondary to nitrate as a nitrogen source.

The deposition rates for both soluble reactive and total soluble phosphorus were considerably higher at Lake Dillon than at Como Creek (3-5X). Since the mechanisms of phosphorus transport through the air are poorly known, no explanation can be offered. However, the observed phosphorus transport values are within the range of values reported for a variety of sites in the literature (Likens et al. 1977). Particulate

Table 43. Loading Rates for Bulk Precipitation at the Main Collecting Station, 1981 and 1982

	1981 (mg/m^2/week)		1982 (mg/m^2/week)	
	Mean	SE	Mean	SE
NO_2-N	0.079	0.008	0.125	0.011
NO_3-N	2.616	0.284	3.724	0.351
NH_4-N	3.479	0.353	1.627	0.098
Total soluble N	18.380	1.776	6.002	0.364
Sol. reactive P	0.732	0.087	0.630	0.123
Total soluble P	1.012	0.143	0.695	0.134
Total particulates	187.0	8.01	152.0	7.54
Particulate P	0.219	0.015	0.178	0.009
HCO_3^-	19.8	8.60	23.4	1.56
H^+	0.055	0.006	0.040	0.006
$SO_4^=$	12.8	0.86	46.7	5.7

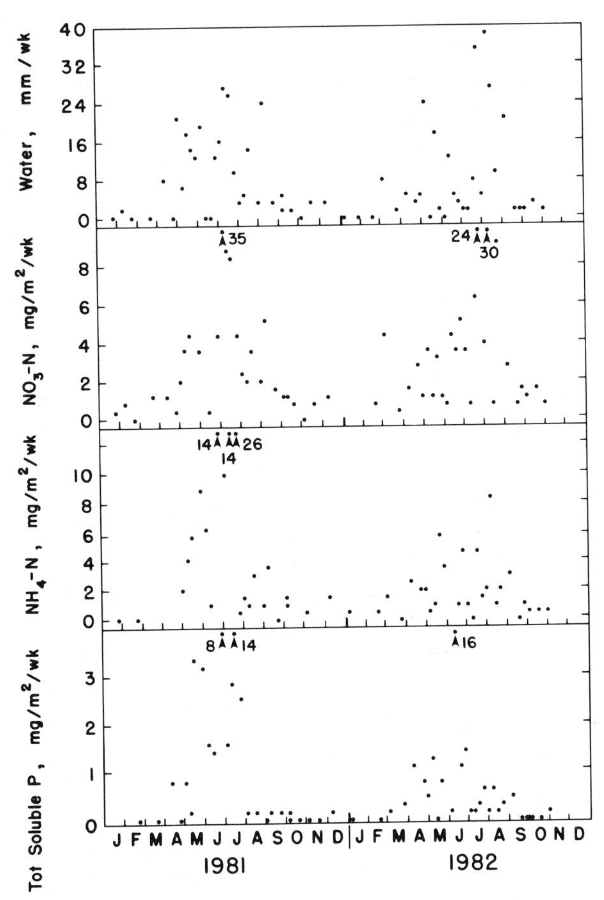

Figure 51. Selected variables for atmospheric deposition near the Snake River WWTP.

phosphorus transport was lower at Lake Dillon than at Como Creek, and this partially offset the higher loading for soluble fractions. Sulfate was a major ion, as would be expected by analogy with the Como Creek studies.

Figure 51 shows time plots of deposition rates for some of the nutrient fractions that have a seasonal pattern. The transport of both nitrate and ammonium is facilitated by moisture, as is evident from a comparison of the amount of rain or snow and the deposition rates for these ions. Total soluble phosphorus showed a pronounced maximum in June of both years. This early summer maximum, although quite high, appeared both years and at more than one station and was thus not an error or a sign of contamination.

Precipitation Chemistry at Other Stations

Phosphorus deposition rates were determined at two additional stations in 1982 as a comparison with the main station located near the Snake River Wastewater Treatment Plant. The first of the comparison stations was located near Frisco. A different type of

Table 44. Summary of Comparison for Phosphorus Loading by Bulk Precipitation at the Main Station and Two Other Stations

Dates of Comparison	Total P Main Station (mg/m²/week)		Comparison Station	Total P Comparison Station (mg/m²/week)	
	Mean	SE		Mean	SE
1 June–17 November	1.354	0.969	Frisco	1.561	0.688
1 July–1 November	0.427	0.065	Raft	0.533	0.099

collector (tube collector, diameter 25 cm, height 1.2 m) was used at this station; the dates of comparison were 1 June to 17 November 1982. The means and standard errors for the main station and the Frisco station over this interval are summarized in Table 44. There was no significant difference in the average loading rates for the two stations ($p > 0.05$). The period of collection incorporated the high phosphorus deposition pulse in June, which results in relatively high standard errors for both stations. However, the pulse appeared simultaneously at the two stations, verifying our conclusion that it was not an artifact.

The second comparison station was situated on the Denver Water Department's raft, which was anchored on the water surface. This comparison station is of special interest because it shows whether a significant amount of phosphorus transport observed at the Snake River station could be accounted for by localized dust in that area. The dates of comparison were 1 July through 1 November. From 1 July through mid-September the raft was anchored in midlake, and from mid-September through October it was anchored at the shoreline. The means for the main station and the station on the raft over these dates are summarized in Table 44 along with their standard errors. The standard errors are much lower in this case because the interval of sampling did not include the period of high phosphorus deposition in June. There was no significant difference in the deposition rates at the two stations.

From the data at the two comparison stations, we have every reason to believe that the main precipitation collection station at the Snake River Wastewater Treatment Plant gave a valid estimate of the nutrient deposition to the surface of Lake Dillon.

13. Total Nutrient Loading of the Lake

The nutrient chemistry data for rivers and effluents must be expressed in terms of transport or loading rates to support an analysis of lake-watershed coupling. Total transport is dealt with in this chapter, which is followed by a dissection of sources and mechanisms of transport in subsequent chapters.

Water Budget

Reliable estimates of total nutrient loading of the lake over the period of study require sound information on the amounts of water reaching the lake by various pathways. If the flow measurements summarized in the previous chapter are accurate, their summation plus the unmeasured flows should be close to the calculated input determined by the Denver Water Department, which is obtained by an indirect method based on measured outflow and change in lake volume. Table 45 summarizes the flows from various sources in 1981 and 1982. The sources of the estimates for the rivers, streams, and effluents are as given in Chapter 12. Two sources of water were estimated without direct measurement: miscellaneous surface runoff and groundwater.

Miscellaneous surface runoff was the contribution from a number of small areas near the lake that drain through watersheds where there was no gauging or discharge measurement. These areas total 4800 ha, or 5.8% of the total land area of the Lake Dillon watershed. Their unmeasured contribution to runoff was estimated in two parts. The first of these is the Meadow Creek drainage, which includes some high-

Table 45. Summary of Water Flow into the Lake in 1981 and 1982

	1981 (Thousands)		1982 (Thousands)	
	m^3/year	Acre-ft/year	m^3/year	Acre-ft/year
Rivers				
Snake	40,100	32.54	81,800	66.35
Blue	49,900	40.45	98,200	79.61
Tenmile	58,000	47.02	104,000	84.31
Subtotal	148,000	120.0	284,000	230.3
Streams not on rivers				
Miner's	2,770	2.25	7,760	6.29
Soda	380	0.31	1,700	1.38
Subtotal	3,150	2.55	9,460	7.67
Effluents				
Frisco	403	0.33	457	0.37
Snake	423	0.33	460	0.37
Subtotal	826	0.66	917	0.74
Diffuse surface drainage	3,366	2.73	10,385	8.42
Precipitation	4,336	3.52	4,057	3.29
Groundwater	8,441	6.90	8,441	6.90
Grand total	168,119	136.4	317,260	257.4

elevation areas. This drainage was considered to be hydrologically similar to Miner's Creek. The runoff was thus set at an amount equal to that of Miner's Creek, with appropriate adjustment for area. The second portion consists of smaller drainages that are low-lying and without steep contours. Because of low accumulation of snowpack in such areas, the water yield was expected to be small. This second portion was assumed to have the water yield of Soda Creek drainage, with appropriate adjustment for area. Any errors in the approximation procedure will obviously not have a very great effect on the water budget of the lake because of the relatively small proportion of the watershed in the miscellaneous surface drainage category.

Groundwater also potentially supplies water, and thus nutrients, to the lake. A complete water budget was constructed on a monthly basis for 1976-1979 from the outflow figures of the Denver Water Department, the inflow data of USGS gauges, measures of precipitation, estimates of evaporation, and an allowance for miscellaneous surface drainage. These terms in the water budget should sum to zero if all estimates were made with good accuracy and if there was no groundwater entry into the lake. Some variance around the zero sum is expected due to inaccuracies of the various estimates and to small miscellaneous factors such as bank storage, but a trend toward greater outflow than inflow would suggest the presence of an unmeasured term, which would have to be groundwater. The residuals over the three years had an average of 575 acre-ft per month. There was no strong seasonal or annual pattern in the size of the residual, and the standard error was about 300 acre-ft per month. This residual may have been affected by the amount of drawdown as well as other factors.

Since the irregularities in the residual cannot be connected to a specific pattern in other variables, we treat the groundwater inflow to the lake as constant at 575 acre-ft per month for both years of the study. Since this amount of flow is very small by comparison with the total flow to the lake, any errors in the estimate within the range of one or two standard errors would not be significant to the overall conclusions of the study.

When all sources of water for the lake were summed (Table 45), the estimated inflow was 136,000 acre-ft for 1981 and 257,000 acre-ft in 1982. This figure can be compared with the independently derived estimate of the Denver Water Department for total input over the same interval. First, however, evaporation must be added to the Denver Water Department estimate, which does not include evaporation. From the averages of several years of study by the Denver Water Department, we estimate evaporation as 5900 acre-ft per year, which gives a corrected Denver Water Department input of 111,000 acre-ft per year for 1981 and 223,000 acre-ft per year for 1982. The deviation of this from the totals based on USGS data as described above is 10-20%, which is more than anticipated. Reexamination of the data did not reveal any potential errors of this magnitude. For nutrient loading computations, we assume that the USGS data for flow are correct, although this assumption is to some degree arbitrary.

Phosphorus Loading

Table 46 summarizes the phosphorus loading of Lake Dillon according to the various pathways of nutrient flow that were recognized in Chapter 12 and in the preceding section on water balance. For rivers, streams, and effluents, the numbers in Table 46 were obtained by multiplying the discharge-weighted average concentrations as reported in Tables 40 and 41 times the time-weighted average discharge and dividing the product by lake area. The estimate of atmospheric deposition was taken directly from Table 43. For miscellaneous surface drainage, the water contribution has already been obtained by approximation as reported in Table 45. The concentration of phosphorus fractions in this water must also be approximated before an estimate of loading can be made. Phosphorus concentrations in the miscellaneous surface drainage category were assumed equal to those observed in Miner's Creek because of parallel land usage, and these concentrations were applied to the discharge approximations to produce the estimate of loading shown in Table 46. A similar approach was taken for groundwater. The groundwater phosphorus concentrations were assumed to be equal to those of the Snake River drainage, a large watershed with mixed land use but lacking point sources. It was also assumed, however, that particulate phosphorus would not be transported through groundwater and this fraction was therefore set to zero. The concentrations were then multiplied by the amount of groundwater as reported in Table 45 to obtain the loading of phosphorus via groundwater. Both the miscellaneous surface drainage and groundwater categories were minor contributors, so the details of these approximations are of no great concern.

More than half of the total phosphorus from all sources entered the lake in soluble form. Table 46 shows that the bulk of total soluble phosphorus was in the soluble reactive fraction rather than the organic phosphorus fraction. The particulate fraction

Table 46. Summary of Phosphorus Loading of the Lake in 1981 and 1982

	Phosphorus (kg/ha/year–1981)					Phosphorus (kg/ha/year–1982)				
	Soluble Reactive P	Total Soluble P	Part. P	Total P	%	Soluble Reactive P	Total Soluble P	Part. P	Total P	%
Rivers										
Snake	0.036	0.060	0.127	0.187	8.7	0.166	0.386	0.423	0.808	22.0
Blue	0.524	0.524	0.206	0.729	33.8	0.316	0.492	0.411	0.903	24.5
Tenmile	0.048	0.156	0.308	0.464	21.5	0.164	0.428	0.459	0.887	24.1
Subtotal	0.608	0.740	0.641	1.380	64.0	0.646	1.306	1.293	2.598	70.6
Streams not on rivers										
Miner's	0.004	0.007	0.013	0.020	0.9	0.017	0.030	0.034	0.064	1.7
Soda	0.002	0.005	0.003	0.008	0.4	0.017	0.026	0.028	0.054	1.5
Subtotal	0.006	0.012	0.016	0.028	1.3	0.034	0.056	0.061	0.117	3.2
Effluents										
Frisco	0.016	0.023	0.005	0.029	1.3	0.170	0.190	0.020	0.210	5.7
Snake	0.001	0.007	0.015	0.022	1.0	0.008	0.013	0.041	0.054	1.5
Subtotal	0.018	0.030	0.020	0.051	2.4	0.178	0.204	0.061	0.265	7.2
Miscellaneous surface drainage	0.011	0.019	0.035	0.054	2.5	0.045	0.080	0.091	0.171	4.7
Precipitation	0.381	0.520	0.114	0.634	29.4	0.328	0.361	0.093	0.454	12.4
Groundwater	0.006	0.010	0.000	0.010	0.5	0.028	0.066	0.000	0.066	1.8
Grand total	1.030	1.331	0.826	2.157	100.1	1.259	2.073	1.599	3.671	99.9

was significant both years and was slightly higher proportionally in 1982 than in 1981, probably because of higher discharge in 1982. The total phosphorus transport in 1981 was about 60% of the total phosphorus transport in 1982.

The rivers contributed about two-thirds of the total phosphorus reaching the lake. The individual rivers made very different relative contributions in the 2 years, however. All were affected by higher discharge in 1982, but not identically. Loading from the Snake River was drastically affected by construction in the river bottom in the second year. Loading from Tenmile Creek was approximately proportional to the difference in flows between the 2 years. The Blue River loading was exaggerated in 1981 by problems at the Breckenridge WWTP. Other differences in sensitivity of the three rivers to changes in the hydrologic conditions between years can be traced to their differing land use patterns; these will be considered further when land use is analyzed.

Precipitation accounted for a large proportion of the residual loading beyond the contributions of the three major rivers. The proportion of total phosphorus loading attributable to precipitation was considerably larger in 1981 than in 1982. This is principally explained by the much greater contributions from surface runoff in 1982, which reduced the overall influence of atmospheric deposition.

The combined contribution of the Frisco and Snake River wastewater treatment plants was only 2.4% in 1981 and 7.2% in 1982. The contribution in 1982 should actually have been lower in view of the much higher phosphorus loading from non-point sources that year, but treatment plant malfunctions produced an exceptionally high effluent contribution on certain dates in 1982. There are two other major wastewater treatment plants in the watershed: Breckenridge and Copper Mountain. The contributions of all four plants will be specified more exactly when the nutrient loadings are broken down in Chapter 15.

Nitrogen Loading

Table 47 summarizes the nitrogen loading of Lake Dillon. The methods of computation and estimation were similar to those used for phosphorus. Particulate nitrogen was not measured in the effluents in 1982 because it was such a small contributor to effluent N. The particulate N column in the table was filled in for 1982 by use of the 1981 data.

The inorganic fraction of nitrogen was slightly larger than the organic fraction for all sources combined. Unlike phosphorus, nitrogen loading was mainly associated with the soluble fractions; particulate nitrogen made a relatively small contribution. The difference between years in total nitrogen loading was very similar on a proportional basis to that observed for phosphorus: the loading for 1982 was almost double that of 1981. For nitrogen, the dominating effect of the three rivers on total loading was even more pronounced than it was for phosphorus. The influence of precipitation on nitrogen loading was relatively small by comparison with phosphorus. The contributions of the two effluents on a percentage basis were slightly larger than for phosphorus, but still relatively minor by comparison with total loading (Figure 52).

Table 47. Summary of Nitrogen Loading of the Lake in 1981 and 1982

	Nitrogen (kg/ha/year–1981)						
	NO_2-N	NO_3-N	NH_4-N	Total Soluble N	Part. N	Total N	%
Rivers							
Snake	0.02	2.86	0.33	5.49	.99	6.49	10.1
Blue	0.37	8.49	6.99	20.61	1.05	21.65	33.8
Tenmile	0.15	9.20	1.13	15.49	2.86	18.36	28.6
Subtotal	0.55	20.55	8.45	41.59	4.90	46.50	72.5
Streams							
Miner's	0.002	0.07	0.03	0.29	0.07	0.36	0.6
Soda	0.00	0.004	0.003	0.03	0.02	0.04	0.1
Subtotal	0.002	0.07	0.03	0.32	0.09	0.41	0.7
Effluents							
Frisco	0.04	0.48	0.84	2.69	0.05	2.74	4.3
Snake	0.04	1.12	0.56	2.15	0.16	2.31	3.6
Subtotal	0.08	1.59	1.40	4.84	0.21	5.05	7.9
Miscellaneous surface drainage[a]	0.005	0.18	0.08	0.79	0.19	0.99	1.5
Precipitation	0.004	1.36	1.81	9.56	1.17[a]	10.72	16.7
Groundwater[a]	0.002	0.23	0.03	0.44	0.000	0.44	0.7
Grand total	0.64	23.98	11.79	57.54	6.56	64.11	100.0

	Nitrogen (kg/ha/year–1982)						
	NO_2–N	NO_3–N	NH_4–N	Total Soluble N	Part. N	Total N	%
Rivers							
Snake	0.09	7.16	0.73	14.33	6.55	20.88	18.4
Blue	0.24	7.98	7.86	28.06	5.36	33.42	29.5
Tenmile	0.86	13.81	7.32	37.05	6.93	43.98	38.8
Subtotal	1.17	28.95	15.91	79.44	18.84	98.28	86.7
Streams							
Miner's	0.008	0.12	0.06	0.59	0.48	1.07	0.9
Soda	0.003	0.08	0.02	0.26	0.13	.38	0.3
Subtotal	0.01	0.20	0.07	0.85	0.61	1.46	1.2
Effluents							
Frisco	0.04	0.42	0.98	2.43	0.05[a]	2.48	2.2
Snake	0.09	0.78	0.67	2.89	0.16[a]	3.05	2.7
Subtotal	0.13	1.20	1.66	5.32	0.21[a]	5.53	4.9
Miscellaneous surface drainage[a]	0.02	0.32	0.16	1.59	1.30	2.89	2.5
Precipitation	0.07	1.140	0.85	3.12	0.95[a]	4.07	3.6
Groundwater[a]	0.01	0.58	0.06	1.16	0.000	1.16	1.0
Grand total	1.41	32.39	18.71	91.48	21.91	113.39	99.9

[a] Estimated.

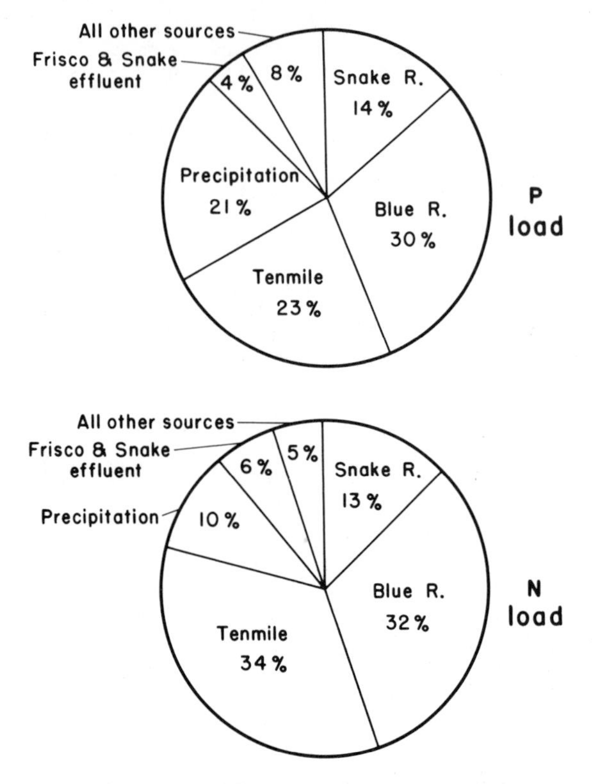

Figure 52. Itemization of sources of P and N loading for Lake Dillon at the points of entry to the lake (average 1981, 1982).

Overview of Phosphorus and Nitrogen Loading

Table 48 expresses the phosphorus and nitrogen loading in several different ways. Because of phosphorus removal at the point sources and low background phosphorus in the watersheds, the molar ratio of nitrogen to phosphorus in the loading was very high: 69 in 1981 and 66 in 1982. Even though this ratio is high, it is only half the

Table 48. Summary of N and P Loading

	Per Unit Lake Area[a]		Entire Lake	
	kg/ha/year	g/m^2/year	kg/year	lb/year
1981				
N	85.3	8.5	85,600	189,000
P	2.9	0.29	2,900	6,300
1982				
N	134.0	13.4	151,400	333,800
P	4.2	0.42	4,800	10,700

[a] Based on mean lake areas of 1004 ha in 1981 and 1130 ha in 1982.

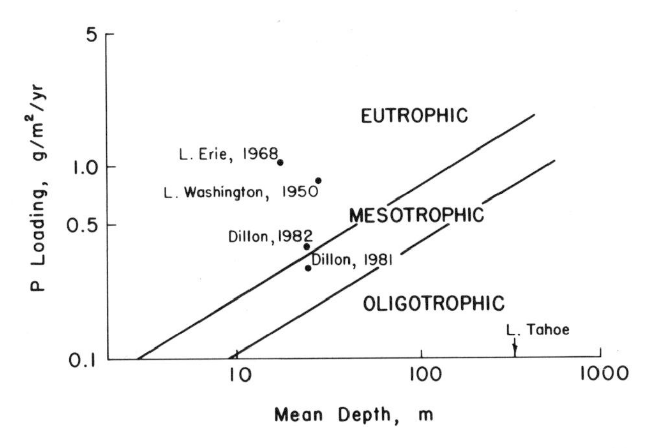

Figure 53. Position of Lake Dillon on the the original Vollenweider diagram (Vollenweider 1968).

average ratio for the lake (Table 20). Phosphorus obviously has a much higher sedimentation rate in the lake than nitrogen. This is explained in large part by the much greater fraction of nitrogen that is in soluble form.

Figure 53 puts the phosphorus loadings into perspective by use of the Vollenweider diagram (Vollenweider 1968), which predicts the trophic status of a lake on the basis of its mean depth and its phosphorus loading. The position of Lake Dillon on the diagram is compared with the position of a small selection of other lakes. The diagram provides only a crude approximation of trophic status because it does not allow the phosphorus loading to be discounted for varying degrees of phosphorus sedimentation and loss through outflow. However, the diagram is in general agreement with the lake trophic indicators. A more detailed treatment of this matter will be given in connection with modelling.

14. Nutrient Export in Relation to Land Use

Nutrient yield of land surface will be considered here in relation to land use. The central information base for this purpose is the representative watershed program, in which nutrient loss was measured for individual small watersheds dominated by particular land uses. The names and locations of the representative watersheds are shown in Figure 54. The watersheds represent eight different land uses: undisturbed (background), roads, interstate highways, residential development on sewer (nonpoint component), urban development on sewer (nonpoint component), residential development on septic, ski areas, and Climax Molybdenum operations. The nutrient yields from these representative watersheds are considered in sequence below. For each of the watershed types except the undisturbed type, the background nutrient yield is subtracted from the total yield and the residual is related quantitatively to the intensity of land use by some index such as the number of persons per unit area or the percentage of affected area. After the nonpoint sources have been treated in this way, consideration is given to the contributions per capita from small and large point sources (wastewater treatment plants).

Water yield enters into many of the nutrient yield relationships developed below for the representative watersheds. The water yields, given as millimeters per year of runoff, are in all instances estimated from discharge measurements. Water yields vary a great deal; the causes of this variation are numerous. There is a pronounced increase in water yield with elevation. Long-term records of snow accumulation also show that, for a given elevation, the western portion of the catchment tends to receive more winter moisture than the rest. Slope, vegetation, and land use also have complex

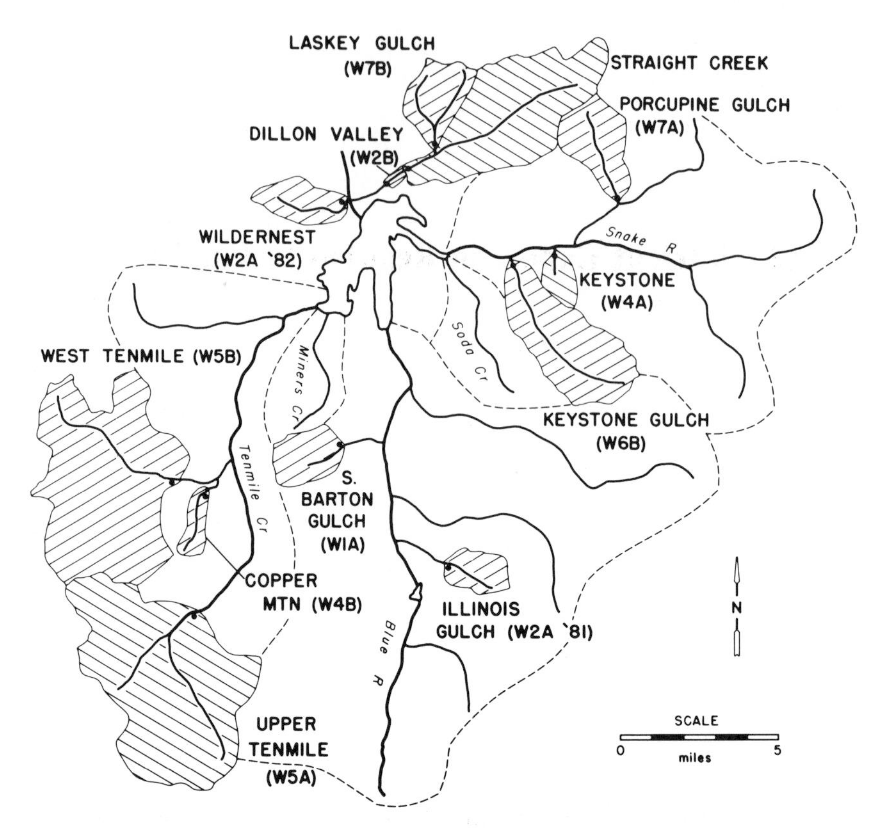

Figure 54. Representative watersheds sampled during the Lake Dillon study.

effects. A certain amount of water is also pumped or diverted in various places (e.g., local irrigation). For 1981, the catchment-wide average runoff at lakeside was 180 mm and in 1982 it was 340 mm, but individual small watersheds scattered widely around these averages.

Undisturbed Watersheds

Porcupine Gulch and Laskey Gulch provided information about nutrient yield from undisturbed areas. Table 49 summarizes the amount of runoff from these watersheds and the phosphorus and nitrogen yield in each of the 2 years. In these and a number of other representative watersheds the particulate nitrogen contribution was computed from the percent carbon in total particulates of the stream by use of a common C:N ratio equal to that of the Snake River, since the amount of particulate N was often below detection. The standard error for phosphorus is reported on the basis of weighted averaging. For nitrogen, the estimation of particulate nitrogen from carbon precluded computation of a standard error, but the ratio of SE to mean would be roughly the same as for phosphorus.

Table 49. Yield of Water, Total P, and Total N from Two Watersheds Representing Background (Undeveloped) Conditions (W7A, W7B), and from Two Watersheds with Roads but Otherwise Essentially Undeveloped (W6B, SR3)

	Water Yield (mm/year)	P yield ($mg/m^2/year$)			N yield ($mg/m^2/year$)	
		Total	SE	% Part.	Total	% Part.
Porcupine Gulch (W7A)						
1981	315	1.55	0.34	17	75	7
1982	541	3.52	0.43	32	126	35
Laskey Gulch (W7B)						
1981	129	0.50	0.10	33	29	12
1982	306	2.77	0.55	33	62	58
Keystone Gulch (W6B)						
1981	108	0.57	0.08	33	14	19
1982	328	3.19	0.44	38	46	41
Upper Snake (SR3)						
1981	240	1.14	0.15	61	64	15
1982	500	4.37	0.55	54	103	31

The phosphorus yields of Porcupine Gulch and Laskey Gulch were close to those reported in the literature for cold temperate areas of similar soil characteristics. For example, Wright (1974) reported 1.5 $mg/m^2/year$ P export for the Boundary Waters Canoe area of Minnesota, and Schindler et al. (1976) reported 5.0 $mg/m^2/year$ for the Experimental Lakes Area, Ontario. The data also show great similarity to data for the Como Creek watershed at a similar elevation in Boulder County (Lewis and Grant 1979).

Both total phosphorus and total nitrogen export showed considerable variation between years and between watersheds. Table 49 suggests, however, that this variation was not random; it was associated with the amount of water yield. As shown by the study of Lewis and Grant (1979) for the Como Creek watershed, the increase in soluble phosphorus yield between a dry year and a wet one is slightly higher than the increase in water yield. A quantitative relationship was sought between the water yield and the nutrient yields reported in Table 49. Since the Como Creek studies suggest that the relationships are likely to be slightly curvilinear for undisturbed watersheds, the following equation was used:

$$Y_n = a Y_w^{\,b}$$

where Y_w equals the annual water yield, Y_n equals the annual nutrient yield (P or N), and a and b are the critical parameters of the relationship between nutrient yield and water yield. The equation was log transformed and then tested by linear regression for fit to the data. The fit was excellent ($p<0.01$). A similar procedure for nitrogen also showed a good fit ($p<0.01$).

Data from watersheds with only gravel roads or dirt roads suggested that the effect of these roads might be sufficiently small that they would not raise the nutrient export much above background. Since this would be an advantage in that it would offer a larger number of watersheds from which the terms relating water and nutrient yield could be determined for background conditions, two watersheds differing from background only by the presence of dirt or gravel roads were examined. These were Keystone Gulch, which was a representative watershed (W6B), and a segment of the upper Snake (SR3), which was sampled as part of the stream survey coverage of the Snake River (Figure 3). The data for these watersheds are shown in Table 49. Keystone Gulch actually has near the watershed crest a small septic field that serves the upper lodge at Keystone ski area, but it is of sufficiently small size that it cannot account for more than 2% of N or P export from Keystone Gulch, so the septic effect was ignored.

When the equations for the two undisturbed watersheds were used to project the yield from the two watersheds with small roads, and the projections were subtracted from the observed yields, two of the nitrogen residuals and one of the phosphorus residuals were slightly negative, suggesting that the yields of the two watersheds with small roads were very near background. The exponent (b) derived for phosphorus yield on the basis of the two undisturbed watersheds was 1.39, and with all four watersheds it was 1.37. The coefficient (a) was higher by 20% for all four watersheds than for the two undisturbed watersheds, but this is a small amount in view of the variation in background caused by differences in vegetation, exposure, elevation, and slope. We therefore conclude that all four watersheds should be treated together, and that the effect of small roads in undeveloped areas is minor enough to merge with background for present purposes. Thus 8 watershed years are available for determination of background yield equations for N and P.

Table 50 summarizes the results of the log-transform regressions used to determine the coefficient and exponent of the background yield equations for the 8 watershed years. The goodness of fit is excellent both for phosphorus and for nitrogen. For phosphorus the exponent is slightly greater than 1, as expected from the Como Creek work. For nitrogen the exponent is lower than for phosphorus. Once again, this is expected for undisturbed watersheds of this type from the Como Creek studies (Lewis and Grant 1979). The Como Creek work shows that nitrate yield typically increases more slowly than water yield, but that particulate N yield increases faster than water

Table 50. Summary of Statistical Information on Empirical Establishment of the Relation $Y_n = aY_w^b$ for Background Yield (Y_w = mm/year Water Yield; Y_n = mg/m^2/year Nutrient Yield)[a]

	Exponent (b)	Standard Error (b)	Coefficient (a)	Equation
Phosphorus[b]	1.372	0.20	0.000782	$Y_n = 0.000782 Y_w^{1.372}$
Nitrogen[b]	1.154	0.17	0.0842	$Y_n = 0.0842 Y_w^{1.154}$

[a] Both equations are highly significant ($p<0.001$).
[b] mg/m^2/year.

yield. The two effects partially cancel, producing an exponent in this instance only slightly greater than 1.0.

Residential Area on Sewer

A number of developments and subdivisions in the Lake Dillon watershed consist of clusters of homes or condominiums served by sewer. These make a point-source contribution, which will be considered at the end of this chapter; they also potentially make a nonpoint contribution to nutrient loading. The nonpoint contribution is associated with the presence of pavement, changes in vegetation and soil cover within the development, fertilizer application, frequent minor earth distrubance, and other miscellaneous factors arising from human habitation. We consider this classification to apply to settlements or developments that have enough vegetative cover to break up runoff and thus slow the transport of materials to streams. This classification does not apply to areas that are very densely settled (more than 80% land coverage by concrete or dwellings). Such densely settled areas are treated below as urban area on sewer.

Several representative watersheds were initially chosen for quantification of the effect of residential development served by sewer. All of these except one proved impossible to use. Some were settled too sparsely and others lacked sufficiently coherent drainage to give good quantitative information. The information is thus based on a single watershed containing the developments of Wildernest and Mesa Cortina, the composite of which we shall call Wildernest. Data are available only for 1982. The data are summarized in Table 51. This watershed has an area of 246 ha, of which 126 ha is taken up by the residential development on sewer. The watershed is occupied by an annual average of 662 persons per day. It is ideal for present purposes in that it is settled densely enough to show any effects rather clearly, but is not so densely settled as to constitute an urban area. Most of the dwellings or groups of dwellings are separated by trees and most areas are unpaved.

Table 51 gives the total water yield and the total phosphorus and nitrogen yields for the watershed. The background, as computed from the background equations, has been subtracted from these to give the net yield attributable to the land use under consideration here. Table 51 expresses this net yield in terms of the use intensity index, which in this case is the number of persons. The net yields per use unit in 1982 were 5.63 g of phosphorus per person per year and 308 g of nitrogen per person per year. These are relatively small in relation to the sewage export per capita, which would be in the vicinity of 400–600 g per year for phosphorus and about five times this amount for nitrogen.

The question arises whether or not the yields shown in Table 51 are dependent on water yield. By analogy with the undeveloped watersheds, it seems certain that there is a positive relationship between nutrient yield and water yield, but this relationship cannot be quantified without more data. It will be assumed hereafter that total phosphorus and total nitrogen yield have a dependence on water yield of the same degree as documented for background yields. With the exponent fixed by this means, the coefficient can be determined algebraically from Table 51. The coefficient for P is 0.00295 and for N is 0.536 to give grams per person per year.

Table 51. Yield of Water and Nutrients from a Watershed Supporting Residential Area on Sewer (1982 Data, Nonpoint Contribution Only), and a Watershed Supporting Urban Development on Sewer (1982 Data, Nonpoint Only)

	Residential on Sewer	Urban on Sewer
	Wildernest (W2A)	Dillon Valley (W2B)
Water yield (mm/year)	246	326
Use intensity (persons/ha)	2.69	6.83
Phosphorus		
Total yield (mg/m²/year)	3.29	14.7
SE	0.40	4.2
% Particulate	29	100
Background (mg/m²/year)	1.49	2.19
Net yield (mg/m²/year)	1.80	2.89[a]
Net yield per use unit (g/person/year)	5.63	4.22
Nitrogen		
Total yield (mg/m²/year)	131	456
% Particulate	39	56
Background (mg/m²/year)	48	67
Net yield (mg/m²/year)	83	218[a]
Net yield per use unit (g/person/year)	308	318

[a] Net yield after subtraction of interstate highway effect as well as background.

Urban Area on Sewer

Although samples could not be taken in such a way as to segregate cleanly any of the cities within the watershed, there was one small watershed segment almost exclusively on sewer that could be treated as an urban area. This watershed segment, which is referred to here as Dillon Valley, is shown in Figure 54. Development is much more dense in Dillon Valley than in Wildernest, which accounts for the difference in classification. The watershed segment has an area of 115 ha. The settled area is 65 ha, of which 62 ha is occupied by dwellings or roads. The area is occupied by a time-weighted average of 786 persons. The segment is sufficiently small and complex hydrologically that discharge could not be estimated easily. We therefore assumed runoff per unit area equal to Straight Creek, into which Dillon Valley drains. Computation of the urban effect also required subtraction of interstate highway effects. The basis of this correction is given in the section on yield from interstate highways. The results are summarized in Table 51.

The total yield of nitrogen and phosphorus for Dillon Valley was much higher per unit area than for residential area on sewer, but this was expected because of the higher density of development. On a per capita basis, the yield above background was similar to the yield for the residential area on sewer. As with the sewered residential area, we assume that the yield of both nitrogen and phosphorus is related to the water yield by an equation incorporating a common exponent (1.372 for P, 1.154 for N).

Once the exponents are thus set, the coefficients of the equations, as determined algebraically from the data in Table 51, are 0.00150 for P and 0.410 for N to give the yield in units of grams per person per year.

Residential Areas on Septic Systems

Representative watersheds for residential areas on septic systems were South Barton Gulch and Illinois Gulch. Data are available both years for South Barton Gulch and in 1981 only for Illinois Gulch. The data are summarized in Table 52.

South Barton Gulch has a moderate density of development; 194 persons are distributed over 817 ha of watershed. Illinois Gulch is much more sparsely settled; it has a time-weighted average of 25 persons distributed over a watershed of 496 ha. The 3 watershed years provided a good range of water yields. Table 52 shows the net nutrient yield from each of the watersheds after subtraction of the background. The net yield is then expressed as per capita phosphorus and nitrogen contribution. The data for nitrogen and for phosphorus suggest that water yield greatly affected nutrient export from these watersheds. This is to be expected, since nutrients stored by septic systems in the soil are carried out in proportion to the amount of water percolating through the soil.

The per capita phosphorus yields were plotted against water yields for each of the 3 years, and showed evidence of a curvilinear relationship. By regression following log transformation it was established that the equation of best fit is $Y_n = 3.44 Y_w^{0.759}$,

Table 52. Yield of Water, Nitrogen, and Phosphorus for Watersheds Containing Residential Septic Systems

	South Barton Gulch (W1A)		Illinois Gulch (W2A/81)
	1981	1982	1981
Water yield (mm/year)	190	522	72
Use intensity (persons/ha)	0.147	0.147	0.050
Phosphorus			
Total yield (mg/m²/year)	4.14	9.83	0.70
SE	1.61	1.13	0.16
% Particulate	74	61	38
Background (mg/m²/year)	1.04	4.19	0.28
Net yield (mg/m²/year)	3.06[a]	5.50[a]	0.42
Net yield per use unit (g/person/year)	208	374	83
Nitrogen			
Total yield (mg/m²/year)	40	397	13
% Particulate	56	72	50
Background (mg/m²/year)	36	115	11.7
Net yield (mg/m²/year)	1.9[a]	275[a]	1.3
Net yield per use unit (g/person/year)	129	18722	257

[a] Contribution of 74 persons on residential sewer has been subtracted as well as background.

where Y_n is yield above background expressed as grams per person per year and Y_w is water yield (mm/year). The fit is good (p=0.04, SE of exponent is 0.10). The exponent is less than 1.0, unlike the exponent for background P yield. The lower exponent implies that higher water yields were accompanied by P yields that were higher by a less than equal proportion. Individual septic fields thus appear to respond in a manner intermediate between that of background sources, whose yield increases faster than water flow, and a true point source, whose yield would not be affected by water flow.

The regression procedure was repeated for nitrogen. The fit with a curvilinear model was poor; fit with a linear model was much better. The resulting equation is $Y_n = 44.3 Y_w - 5205$, where Y_n is yield of N as grams per person per year and Y_w is millimeters per year water yield. The standard error for the slope is 11.8 and the significance of the relation is p=0.08. The cause of the negative intercept is clear from Table 52: N yield above background becomes vanishingly small at water yields well above 0. This is explained by biological uptake and retention up to a certain flushing threshold, by denitrification, or by a combination of these.

Ski Slopes

The water and nutrient yields from Keystone and Copper Mountain ski slopes (including lodges, trails, and lifts) were quantified in 1981 and 1982. Contributions from

Table 53. Yield of Water and Nutrients from Watersheds Supporting Ski Slopes (Nonpoint Contributions Only)

	Keystone (W4A)		Copper Mountain (W4B)	
	1981	1982	1981	1982
Water yield (mm/year)	98	195	348	329
Use intensity (% open)	40	40	42	42
Phosphorus				
Total yield (mg/m^2/year)	1.94	5.51	3.56	6.29
SE	0.41	1.30	0.99	1.49
% Particulates	49	67	44	47
Background (mg/m^2/year)	0.42	1.08	2.40	2.22
Net yield (mg/m^2/year)	1.52	4.43	1.16	4.07
Net yield per use unit (mg/m^2 used/year)	3.84	11.20	2.40	8.43
Nitrogen				
Total yield (mg/m^2/year)	27	75	74	88
% Particulate	59	53	27	37
Background (mg/m^2/year)	17	37	72	67
Net yield (mg/m^2/year)	10	38	2	21
Net yield per use unit (mg/m^2 used/year)	25	96	4	44

the associated residential areas and point sources are not considered here. Table 53 summarizes the data. The yield of phosphorus was well above background. The nitrogen yield was also above background, but less markedly so than phosphorus yield. The percentage of each watershed accounted for by cleared areas was used as a measure of use intensity; for both resorts about 40% of the area is cleared. The number of skiers would probably serve equally well as an intensity measure. In 1981 and 1982 Copper Mountain and Keystone had between 500,000 and 700,000 skier days per year.

The water yields from Copper Mountain were higher than those for Keystone; this is explained partly by elevation and partly by location along the east-west moisture gradient. There is no indication in Table 53 of the close relation between water and nutrient yield documented for most other land uses. The explanation is not obvious, since the mechanisms of nutrient release are complex. Presence of the trails and slopes undoubtedly facilitates transport, even without human activity, especially since ski trails and slopes must be cleared parallel to the gradient. Possibly more important, however, is mechanical damage to the ground surface when snowpack is minimal. There is also a certain amount of spring fertilizer application on the slopes. Also, yields at either Keystone or Copper Mountain could have been augmented by construction on or near the slopes.

Since the yield from ski slopes does not show any indication of sensitivity to water yield, we use a mean. The mean for the 4 watershed years in Table 53 is 6.5 mg per m^2 of cleared area per year above background for P and 42 for N.

Interstate Highways

The effects of interstate highways were isolated in the upper segments of Straight Creek and West Tenmile Creek. Table 54 summarizes the data and expresses yield above background in relation to area of roadway plus right of way. For phosphorus, the yield was unexpectedly high in both watersheds. The relation of P yield to water yield was tested for fit to a function of the form already used for other nutrient sources. The fit was reasonably good, and resulted in the equation $Y_n = 0.00209 Y_w^{1.799}$, where Y_n is nutrient yield per square meter of highway plus right of way and Y_w is water yield (mm/year). The slope has a standard error of 0.53 and p=0.09 (ideally p should be below 0.05, but the power of the test is low with N=3). The cause of the high yield is unclear. It is probably due in part to major disturbance of vegetative cover. Also the materials added to roads in the winter may contribute, as does the fertilization associated with revegetation programs now in progress.

The nitrogen yield associated with roads was low by comparison with other sources and less consistent than for phosphorus, as shown by Table 54. The N yield above background for West Tenmile was negative one year. This is quite possible due to excess of uptake or denitrification over yield, but presents some difficulties in prediction. For purposes of curve fitting, the two yields from West Tenmile were averaged, giving a low positive yield. The fit to an exponential equation was then reasonably good: $Y_n = 1.71 Y_w^{1.13}$, SE slope = 0.40, p=0.11, where the units are the same as those for the phosphorus equation.

Table 54. Yield of Water, N, and P from Watersheds with Interstate Highways but Little Else

	Straight Creek (W2B2)	West Tenmile (W5B)	
	1982	1981	1982
Water yield (mm/year)	326	522	662
Use intensity (% roads)	2.97	0.78	0.78
Phosphorus			
Total yield ($mg/m^2/year$)	4.09	5.38	7.19
SE	0.84	1.76	0.67
% Particulate	60	47	50
Background ($mg/m^2/year$)	2.19	3.80	5.51
Net yield ($mg/m^2/year$)	1.90	1.57[a]	1.67[a]
Net yield per use unit (mg/m^2 used/year)	64	200	213
Nitrogen			
Total yield ($mg/m^2/year$)	102	181	123
% Particulate	61	18	36
Background ($mg/m^2/year$)	68	115	151
Net yield ($mg/m^2/year$)	34	66[a]	−28[a]
Net yield per use unit (mg/m^2 used/year)	1146	8426	−3574

[a] A small contribution from Vail pass public toilet septic field has been subtracted.

Mining

The Climax Molybdenum mining operation is a unique mixture of nutrient sources and must therefore be treated separately from other land uses. Water from Climax Molybdenum has a high suspended and dissolved solids content. The water is retained in tailings ponds, where sedimentation occurs, and the ponds are purged during high water. The ponds receive not only industrial wastewater but also domestic effluent derived from the toilets and kitchens of the plant. The domestic effluent is not treated, except by sedimentation in the ponds.

Table 55 summarizes the water and nutrient yields from upper Tenmile Creek where the active mines are located. There is little else in this watershed except old mines and State Highway 91. Roads, tailings, and bare areas around the plant account for 753 ha of the total 6447 ha. Virtually all of the nutrient yield above background can thus probably be attributed to Climax Molybdenum.

The P yields were above background and the N yields were even more so in 1981 and 1982. Neither seems related to water yield. Unlike most watersheds, the total nutrient yields for upper Tenmile were higher in 1981 than 1982. In 1981 the time-averaged work force at Climax Molybdenum was 2650 persons and in 1982 it was 1100 persons. The yield of both N and P thus seems related to size of work force. Table 55 expresses the yields both years on a per capita basis; the yield per capita was

Table 55. Yield of Water, N, and P for the Mining Area on Upper Tenmile Creek[a]

	Upper Tenmile (W5A)	
	1981	1982
Water yield (mm/year)	435	513
Use intensity (persons/ha)	0.41	0.17
Phosphorus		
Total yield ($mg/m^2/year$)	7.07	5.84
SE	1.55	0.64
% Particulate	20	58
Background ($mg/m^2/year$)	3.26	4.09
Net yield ($mg/m^2/year$)	3.81	1.75
Net yield per use unit (g/person/year)	93	103
Nitrogen		
Total yield ($mg/m^2/year$)	821	491
% Particulate	5	15
Background ($mg/m^2/year$)	93	113
Net yield ($mg/m^2/year$)	728	378
Net yield per use unit (g/person/year)	17,710	22,150

[a] Use intensity is based on number of persons working at Climax Molybdenum.

almost identical for the 2 years. We will thus treat N and P yield at Climax Molybdenum as a function of work force size.

The P yield per capita at Climax Molybdenum suggests that the P output above background is almost entirely a sewage contribution and not an industrial effect, since the yield per capita is in the range of what would be expected from the domestic output of the workforce. The ponds appear to be effective in reducing particulate P losses that would result from earth disturbance. N export is very high, suggesting major augmentation by industrial activity. Ammonium nitrate is used in blasting and may be the source of this augmentation. Revegetation may also contribute.

Point Sources

There are two types of point sources in the Lake Dillon catchment: large sources with tertiary treatment and small sources with secondary treatment. We will establish here a contribution per capita for each category. Large point sources with tertiary treatment include the Snake River WWTP and Frisco WWTP, whose effluents were sampled directly. The results for these two have already been reported in Chapter 13, but have not yet been converted to a per capita basis. In addition, data can be obtained for the Breckenridge WWTP as the difference between stations BR1 and BR2 on the Blue River (just downstream and just upstream of the Breckenridge WWTP outfall: Figure 3). These data are summarized in Table 56. The Copper Mountain WWTP, designed as a tertiary plant, was in a transition from secondary to tertiary during the course of

Table 56. Yield of P and N from Three WWTP Plants on a Per Capita Basis[a]

	Persons Served[b]	P (g/person/year)		N (g/person/year)	
		1981	1982	1981	1982
Frisco WWTP	3300	11.7	84.8	1108	955
Snake River WWTP	3230	9.1	22.3	1003	1260
Breckenridge WWTP	6700	113.0	44.0	1378	1479

[a] All three plants practice tertiary P removal.
[b] Annual average persons per day.

the study and will therefore not be considered here as representative of tertiary treatment.

Table 56 shows the effectiveness of the tertiary treatment, since all of the per capita P yields are well below the raw per capita contributions coming to point sources

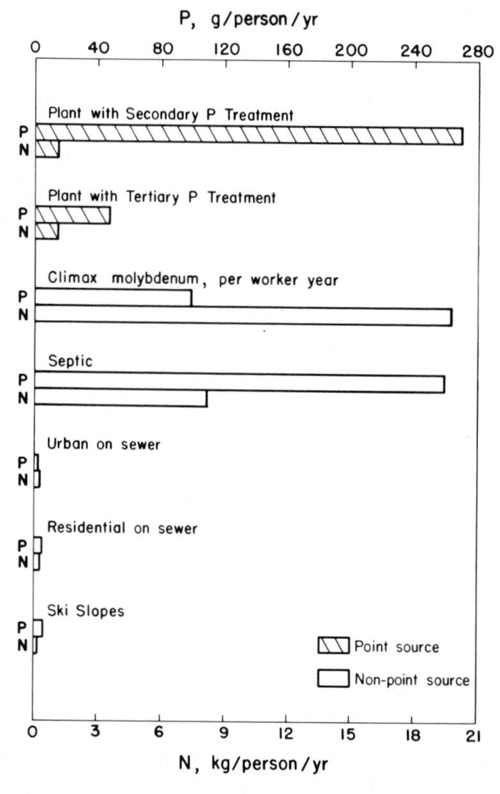

Figure 55. Nutrient yield from various sources expressed on a per capita basis, assuming 300 mm runoff. Sources that cannot be expressed on a per capita basis are not shown.

Table 57. Average Point Source Yields on a Per Capita Basis

	Grams per person per year
Phosphorus	
Secondary treatment	270
Tertiary treatment	47
Nitrogen	
All treatment	1197

(400-600 g/person/year). At the same time, there was considerable variation in P yield per capita between years, even at the same plant. This is explained entirely by breakdown or shutdown of tertiary treatment, which occurred at all plants on one or more occasions over the course of the study. Although it could be argued that tertiary sources should be represented by their yields when treatment facilities are operating

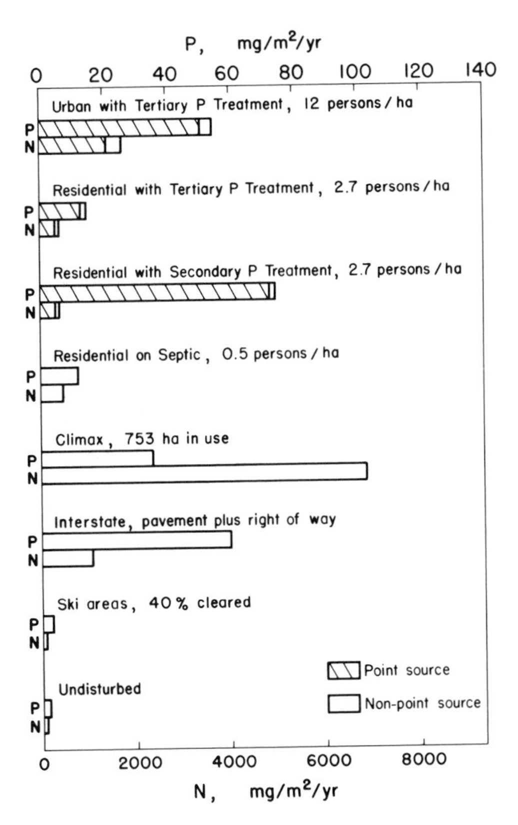

Figure 56. Nutrient yield from various sources expressed on an areal basis, assuming runoff of 300 mm.

properly, which would give yields about half as great as the observed average (Table 57), it is probably realistic to assume that the plants will be impaired for a certain amount of time each year.

Nitrogen yields from the point sources were much more uniform, since they were not affected as much by variations in plant operation. The Breckenridge plant practices ammonium removal, but this does not affect total N yield per capita (Tables 56 and 57).

The P contribution from small point sources (package plants) subject to secondary treatment was estimated from data taken upstream and downstream of such sources along the Blue River above Goose Pasture Tarn. The results are reported in Table 57.

Overview of Nutrient Yields

Export of N and P from various sources is expressed in Figure 55 on a per capita basis whenever this is meaningful. In Figure 56 the export rates are recast on an area basis, assuming runoff is near that of an average year. Among the point sources, the potency of sources without tertiary treatment is striking. For nonpoint sources, septic systems stand out, as expected, as do interstate highways, which might not have been expected.

15. Separation of Nutrient Sources within the Watershed

Chemistry and discharge measurements were taken at a number of points along each one of the three major rivers (Figure 3). Each of the river drainages was thus divided into a number of watershed segments. If the land use within any given segment is known quantitatively, the equations that have been developed in the previous chapter for nutrient yield associated with various land uses should provide a prediction of the nutrient yield from that segment. Three important factors may cause deviations between the observed and predicted nutrient yields from any given watershed segment: random estimation errors, unexpected sources, and storage.

The observed yields for a segment are based on chemistry and discharge measurements, both of which are subject to analytical error. We may also include under the heading of error variance the confidence limits around any one of the curves used to predict nutrient yield from land use. These confidence limits reflect not only error variance in the original data from which the curves were developed but also a certain amount of individuality in watersheds with respect to soils, vegetation, exposure, slope, and elevation. If these sources of error were the only cause of deviations between predicted and observed, we would expect to see a random assortment of positive and negative deviations between the observed and predicted values.

The second cause of deviation between observed and expected yields for individual segments is the presence of nutrient sources that are either unquantifiable or unknown. For the Dillon watershed in 1981 and 1982, there was one instance where such a source proved to be important, as described below in the analysis of nutrient sources of the Snake River drainage.

The third factor affecting agreement between observed and predicted yields is a storage effect associated with areas of low relief in the river bottoms. This has not been incorporated into the equations used for prediction. The equations were developed on the basis of small watersheds off the main channel where individual land uses could be isolated. When nutrients flow to lower elevation, and particularly into a much broader stream channel incorporating low-lying areas and wetlands, significant amounts of nutrient can be stored in years of low flow and purged in years of high flow. Mechanisms of phosphorus storage include sedimentation of particulate phosphorus, biological uptake of soluble phosphorus, and ligation or inorganic precipitation of soluble phosphorus. For nitrogen, the possibilities are more complex, including all of the possibilities for phosphorus plus denitrification, by which oxidized nitrogen species are converted to nitrogen gas, which is subsequently released into the atmosphere. All of these phenomena are most likely in the river bottoms and associated flood plains and wetlands.

Nutrients lost by sedimentation and ligation or uptake in years of low water may be returned to the open channel and thus enter the lake during a year of high water when the physical forces at work in the river bottoms are sufficient to move large amounts of accumulated materials. Thus the river bottoms can act as a storage site that retains a portion of the nutrient yield in dry years and gives back much or all of this retained portion in wet years. These storage and release phenomena are most correctly regarded as temporary changes in the watershed nutrient inventory from one year to the next, and will be referred to below as the "inventory change." If inventory change is important in determining the yield of nutrients to Lake Dillon, we should see in dry years a consistently lower ratio of observed to predicted yield in river segments incorporating areas where the river water moves through wetlands, gravel beds, or ponds. If much or all of the storage (increase in inventory) is given up in a wet year (decrease in inventory), we would expect a consistently higher ratio of observed to predicted yields from these segments in a wet year than in a dry one. If nutrient inventory change is shown by the data from individual stream segments, then additional equations must be developed to account for year-to-year inventory change in the catchment as a whole.

Snake River Drainage

The Snake River drainage was divided into four segments as shown in Figure 57. Histograms of the observed and predicted yields are superimposed on each one of the stream segments in Figure 57. Phosphorus yields are presented on the left side of the river channels and nitrogen yields on the right side of the river channels. For each year and for each segment, one histogram bar shows the observed yield from the segment, and the adjacent bar shows the predicted yield. The scales for the yields are all the same for a given nutrient, but differ between the two nutrients because of the larger yields of nitrogen.

For phosphorus the agreement between predicted and observed yields from individual stream segments is generally very good. One notable exception is the phosphorus yield in 1982 from segment SR2, which incorporates Keystone Gulch, Keystone Ski Area, and Keystone Development. The observed phosphorus yield from

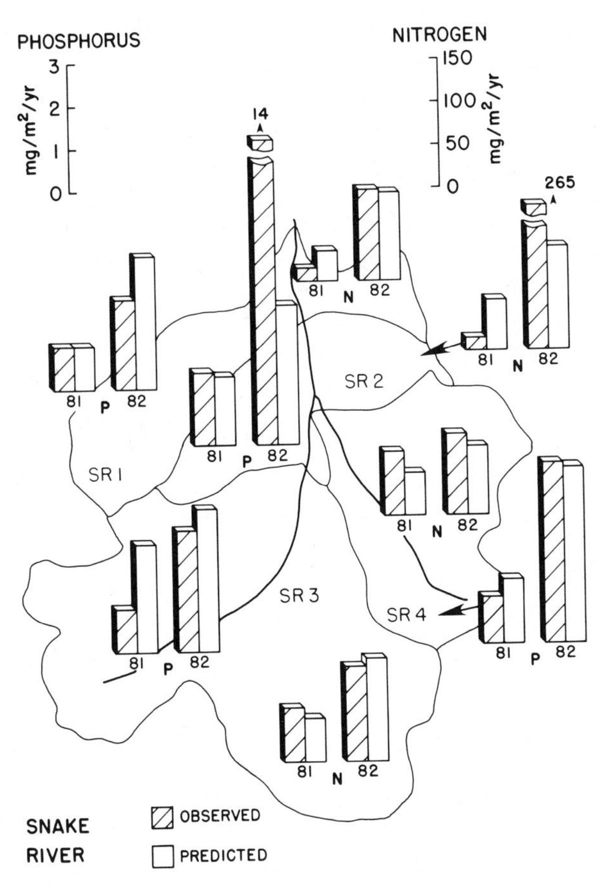

Figure 57. Observed and predicted yields from segments of the Snake River watershed. Phosphorus is on the left of the stream channel and nitrogen is on the right of the stream channel in any given segment.

this segment was extraordinarily high in 1982; it was far above any expected random deviations from the predicted. This exceptionally high observed phosphorus yield is explained by construction activities. Keystone Ski Area installed six 300-hp pumps in the river bed during 1982 and constructed an associated pipeline to carry the water from these pumps to ski slopes, where it was to be used for making snow. There was construction at several other locations in various watersheds during the course of the study, but it did not produce such extraordinary yields of phosphorus. The especially high yield associated with the Keystone construction is undoubtedly accounted for by work in or near the river bed, where water movement greatly facilitated the transport of phosphorus-bearing materials loosened by the construction activities. Since the prediction equations do not take into account any special activities, the gap between observed and predicted P of 11 mg/m^2/year can be explained by the construction. Total 1982 transport due to the construction in this segment was 295 kg of phosphorus. This was approximately 7% of the runoff yield of phosphorus for the entire Lake

Dillon watershed during 1982. Expressed in another way, it was equivalent to the annual yield of phosphorus from a tertiary point source serving 6300 persons, assuming efficiency of point source operation as shown in Table 57.

There is little evidence for significant valley bottom inventory changes affecting phosphorus yield in the four Snake River segments. The inventory change effect is unlikely in SR3 and SR4 because of the steep relief in these segments. The effect is most likely in segment SR2, where it would be obscured by the large construction effect in 1982.

Predictions for nitrogen are also in good agreement with the observed values with the exception of the 1982 value for segment SR2. Once again, we associate the extraordinarily high yield from SR2 in 1982 with the Keystone construction activities.

Blue River Drainage

The Blue River drainage was originally divided into seven segments (BR1–BR7). Figure 58 shows the segment yields based on a total of four segments. Segment BR1

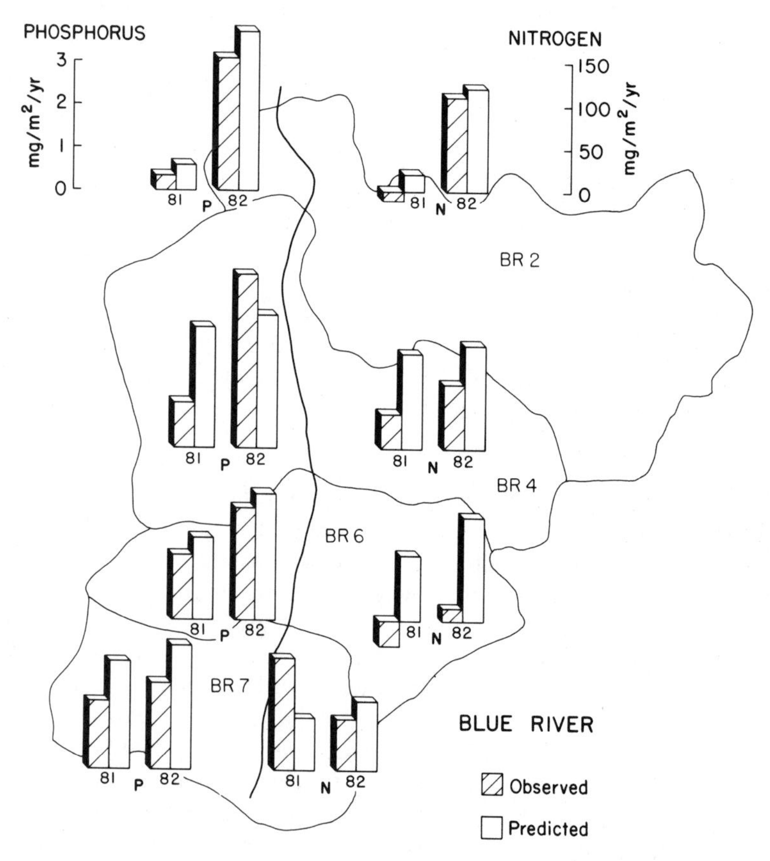

Figure 58. Observed and predicted yields from segments of the Blue River watershed. Phosphorus is on the left of the stream channels, nitrogen on the right.

has been omitted because this small segment isolates only the Breckenridge Wastewater Treatment Plant, whose yield has already been discussed in the previous chapter. Some of the smaller segments between BR2 and BR7 have been combined for present purposes because a considerable land area is required to show a quantifiable effect on the chemistry and discharge of a river as large as the Blue River. Combination of segments was also necessitated by diffuse drainage in the vicinity of Breckenridge, which made impossible any reliable determinations of discharge there. BR2 in Figure 58 includes the original BR2 and BR3; BR4 in Figure 58 includes original BR4 and BR5.

There were no extraordinary departures of observed from predicted phosphorus yield in the Blue River drainage. The Blue River drainage is a good place to look for valley bottom inventory change because the main channel, even as high as BR7, incorporates extensive wetlands, small ponds, and accumulated old tailings in the river bed. In every segment of the Blue River drainage, the ratio of observed to predicted phosphorus was lower in 1981 than in 1982. This is explained by the inventory change effect, which leads to the entrapment of a certain amount of the yield in dry years, and a release of a portion of this material during wet years.

In BR7, the nutrient sources drain through extensive areas of low relief. The predictions were consistently above the observed yields in this portion of the watershed for phosphorus, and the ratio of observed to predicted was higher in 1982 than in 1981. The data thus suggest that significant retention occurred even in the wetter year of 1982, although it was especially pronounced in 1981. The same was true of BR6, which incorporates Goose Pasture Tarn as well as an extensive, diffusely drained area above the Tarn.

Segment BR4 is very different from BR7 and BR6. The river bottom has been extensively modified, both by mining and by settlement around the town of Breckenridge. The phosphorus data show phosphorus accumulation in low-lying areas in 1981 and a strong purging of the stored material in 1982. The behavior of segment BR2 was more similar to that of BR6 than BR4.

There was considerable difference between the yield patterns of nitrogen and phosphorus in the Blue River drainage. For several reasons, phosphorus and nitrogen cannot be assumed to behave similarly in any given watershed segment. First, a great deal of phosphorus is transported in particulate form, whereas the portion of nitrogen transported in particulate form is generally much smaller. Thus the factors that lead to entrapment of significant amounts of particulate phosphorus in a given watershed segment will not necessarily have a similarly strong effect on total nitrogen yield from that segment. Furthermore, nitrogen yields can be affected by denitrification losses. Denitrification is especially likely in low-lying areas with diffuse drainage where anoxia can occur. Finally, the biological uptake of phosphorus and nitrogen is likely to differ.

The ratio of observed to predicted nitrogen yield for individual segments was higher in 1982 than in 1981 for all segments except BR7. The data thus indicate operation of the inventory change effect that was observed for phosphorus. However, it is clear that other factors complicated the yield of nitrogen from most of the segments. Negative yields (i.e., N sources plus incoming N greater than outgoing N) were observed for segments BR2 and BR6 in 1981. Yields in segment BR6 were exceptionally low even in 1982 by comparison with the prediction. The major deviations of segment BR6 from the predictions are almost certainly connected with losses of nitrogen associated with

Goose Pasture Tarn. Concentrations of nitrogen above and below the Tarn indicate that it is a major nitrogen sink. Nitrogen loss in the Tarn is probably due to a combination of uptake by weedbeds, sedimentation, and denitrification in the mud.

Tenmile Creek Drainage

The data for Tenmile Creek are shown in Figure 59. Smaller segments have been combined for reasons similar to those mentioned for the Blue River. The upper segment is the site of Climax Molybdenum, for which predictions are good. The middle segment (TM4) has numerous dispersed sources of nutrients, and is also the discharge point for Copper Mountain Wastewater Treatment Plant. The presence of the wastewater treatment plant complicates the interpretation of the data. The Copper Mountain plant was doing secondary treatment during almost all of 1981, and the effluent was

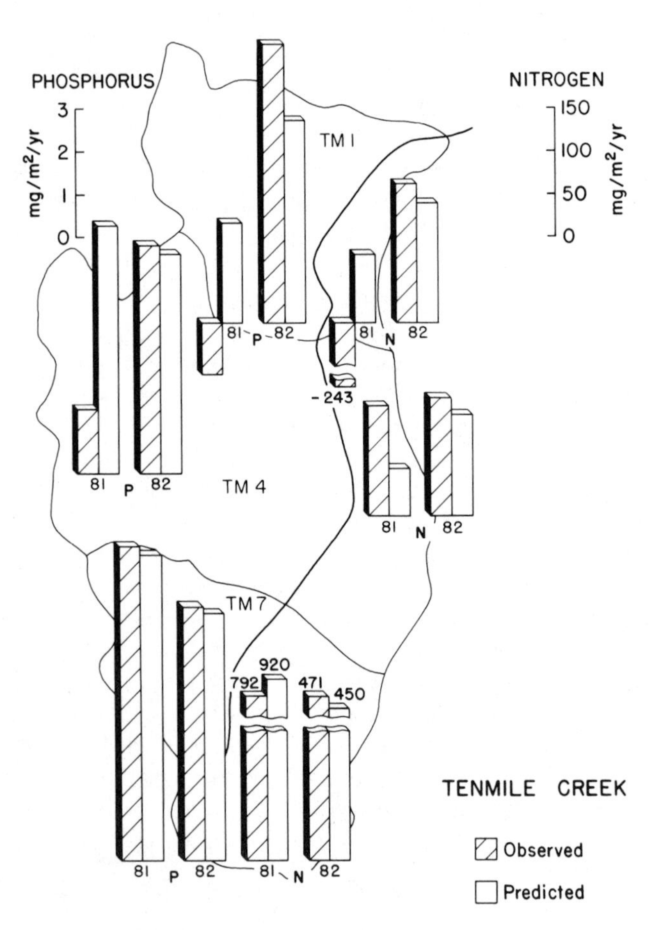

Figure 59. Observed and predicted yields from segments of the Tenmile Creek watershed. Phosphorus is on the left of the stream channels, nitrogen on the right.

for the most part pumped into a pond instead of into the river directly. Tertiary treatment was in force, but was considered not fully effective according to plant operators, during the first 10 months of 1982, after which secondary treatment was resumed for the months of November and December. The presence of the wastewater treatment plant and the problems associated with its new treatment practices make the estimates of other factors difficult because the plant introduced much more irregular variation than would be expected in other watershed segments. There was major overestimation of phosphorus for TM4 in 1981 but not in 1982. The 1981 overestimate is probably due to storage in the pond and in the river bottom in the lower half of the segment; storage would be very likely in the poorly drained areas known as Curtain Ponds and Wheeler Flats. However, segregation of this storage effect is more difficult in TM4 than in any other location because of the changing operations of the point source.

Segment TM1 incorporates a large portion of the town of Frisco. Since the town is located just adjacent to the lake, the drainage of Tenmile Creek near its mouth is modified in a major way by the presence of the town. There is a large area of flat relief in the vicinity of Frisco. Probably because of this flat area, the phosphorus and nitrogen yield data for TM1 show evidence of inventory change between the dry and wet years. Both phosphorus and nitrogen yields for 1981 were negative, indicating that the areas of low relief within this drainage stored not only some of the yield from segment TM1 but also some of the yield of upstream segments. However, the observed storage in 1981 was almost exactly compensated by excess of observed over predicted yield in 1982, indicating a purging effect. The purging effect was less pronounced for nitrogen, possibly because of permanent nitrogen losses to denitrification or biological uptake.

Relationship between Runoff and Nutrient Storage

Since the data show that nutrients originating at various points in the watershed are stored in the river valleys during years of low runoff and purged from these areas during years of high runoff, the yield equations cannot predict total output from the three rivers until they have been coupled with an inventory change function. An estimate can be made of the function from the data at hand. Inventory change will be positive in those years when part of the yield from areas of steeper relief accumulates in areas of low relief and will be negative when part of this stored inventory is released during wetter years. For nitrogen, the function will be considered to include any effects of denitrification.

The first step in estimating the inventory change function is to sum up the observed and predicted total yields from runoff for each of the 2 years of the study. The lakeside point sources (Frisco, Snake River, and Breckenridge wastewater treatment plants) are excluded from the summation, since they are so close to the lake that the nutrients they give off are not subject to storage. The summations are corrected for major discrepancies from the predicted values that are known not to be connected with storage. In the 2 years of record, the only example of this is the boost of yield in the Snake River watershed during 1982 caused by construction in the vicinity of Keystone. After this correction, the remaining discrepancy between observed and predicted yields from all sources combined is assumed to be due to inventory change.

Since inventory change is dependent on amount of runoff, we postulate a decline from positive to negative inventory change as runoff goes from the lowest to the highest annual values. A corollary is that the average change in inventory over a number of years is not very great. Even if the watershed is aggrading, as it may be with net accumulation of vegetation and filling in of excavated areas, the resulting departure from equilibrium over a number of years would amount to a relatively small percentage of nitrogen and phosphorus flux (cf. Bormann and Likens 1979).

Figure 60 shows the apparent change in nutrient inventory for the two years of record for both nitrogen and phosphorus. The nutrient inventory change is expressed as a percentage of watershed overland nutrient yield, excluding lakeside WWTP sources. The percentages were calculated as follows:

$$W_n = \frac{(Y_{n,o} - Y_{n,e})}{Y_{n,o}} \times 100$$

where W_n is nutrient inventory change (%), $Y_{n,o}$ is observed overland yield, and $Y_{n,e}$ is the expected overland yield from summation of all sources with the assumption that inventory change is zero. Figure 60 shows the relationship of W_n to discharge. The two points (1981, 1982) for each of the two nutrients are joined by straight lines. Although the form of the relationship is not necessarily linear, numerous additional points would be required to show curvature. Unless there is very marked curvature, the line will provide reasonable approximations.

The watershed appears to be aggrading with respect to phosphorus, since an average water year is accompanied by a 10–15% retention of yield in valley bottoms. Nitrogen also shows some net retention in an average year, but less than phosphorus. The equations for the lines, as indicated in Figure 60, will be used to represent the expected change in inventory in any particular year.

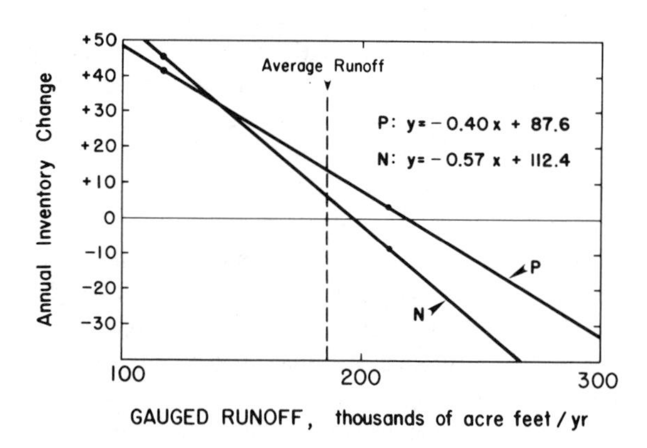

Figure 60. Plot of the inventory change functions for nitrogen and phosphorus. Positive inventory change indicates storage. Inventory change is expressed as a percentage of the summed runoff nutrient sources.

Table 58. Comparison of Predicted and Observed N and P Loads for the 2 Years of Study

	P load (kg/year)		N load (kg/year)	
Year	Observed	Predicted	Observed	Predicted
1981	2900	2700	85,600	83,500
1982	4800	4600	151,400	154,300

Comparison of Observed and Predicted Total P and N Loading

The land use equations, point source equations, and inventory change function lead to predictions of total runoff yield. All identifiable sources are dealt with by these equations except major construction in or near stream beds, such as that observed near Keystone in 1982. Although the frequency of this activity is largely unpredictable, it will surely be above zero. As an approximation of the effect, we use the average increase in yield over the 2 years of record expressed as percent of other runoff. This boosts the P prediction by 3.5% and the N prediction by 2.2%.

When the equations for yield from all watershed sources are used, and a mean figure is used for contributions from precipitation and groundwater, the result is an overall predicted loading of N and P for the lake in 1981 and 1982. These predicted loadings are shown in Table 58, which also shows the observed loading for comparison. The observed and predicted values agree very well (<10% difference). Deviations of predicted from observed are caused mostly by the irregular nature of failures in tertiary treatment plants, and to some extent by irregular variation in other sources.

Itemization of Lake Nutrient Sources for 1981 and 1982

In the chapter dealing with total nutrient loading of the lake, nutrient sources were itemized according to the major parthways by which they entered the lake (Tables 46 and 47). On the basis of the nutrient yield equations, nutrient sources can be further broken down according to land use. In making this breakdown, we use the values reported in Tables 46 and 47 for precipitation, groundwater, and effluent contributions from the Frisco and Snake River plants. The contributions of the Breckenridge Wastewater Treatment Plant are taken from the analysis of watershed segment BR1 as reported in Table 56. The contributions of all other sources are obtained by application of the yield equations and the inventory change equations to the land use data and water yield data for 1981 and 1982. Since the ability of these equations to predict observed yield accurately for 1981 and 1982 has already been demonstrated, it is expected that the sum of individual sources obtained by this means will be very close to, but not exactly equal to, the observed loading.

Table 59 shows the complete breakdown of sources. The table gives the kilograms of nitrogen and phosphorus from each source reaching the lake in 1981 and 1982, and also expresses these as a fraction of total annual loading. Although data for both years

Table 59. Breakdown of Nutrient Contributions to Dillon in 1981 and 1982

	Phosphorus					Nitrogen				
	1981		1982			1981		1982		
	kg	%	kg	%	Mean %	kg	%	kg	%	Mean %
Tertiary outfalls										
Frisco	39	1.3	280	6.1	3.7	3656	4.2	3152	2.1	3.2
Snake River	29	1.0	72	1.6	1.3	3240	3.8	4070	2.7	3.2
Breckenridge	757	25.3	295	6.4	15.9	9233	10.7	9909	6.5	8.6
Copper Mountain	—	—	41	0.9	0.5	—	—	1893	1.2	0.6
Subtotal	825	27.6	688	14.9	21.3	16129	18.7	19024	12.5	15.6
Secondary outfalls										
Package plants	49	1.6	49	1.1	1.3	198	0.2	198	0.1	0.2
Copper Mountain	322	10.8	47	1.0	5.9	1305	1.5	378	0.2	0.8
Subtotal	371	12.4	96	2.1	7.2	1503	1.7	576	0.4	1.0
Climax molybdenum	152	5.1	104	2.2	3.4	28583	33.0	23736	15.6	24.3
Keystone construction	—	—	295	6.4	3.2	—	—	3883	2.6	1.3
Dispersed nonpoint sources										
Residential areas (sewered)	9	0.3	32	0.7	0.5	487	0.6	1901	1.3	1.0
Urban areas (sewered)	9	0.3	31	0.7	0.5	761	0.9	2702	1.8	1.4
Septic	161	5.4	383	8.3	6.9	6713	7.8	21750	14.3	11.1
Interstate highway	21	0.7	132	2.9	1.8	461	0.5	2198	1.4	1.0
Ski slopes	36	1.2	59	1.3	1.3	214	0.2	428	0.3	0.2
Background runoff	545	18.2	2105	45.6	31.8	16577	19.2	69015	45.4	32.3
Subtotal	781	26.1	2742	59.4	42.8	25213	29.2	97994	64.4	46.8
Precipitation	846	28.2	606	13.1	20.7	14311	16.6	5433	3.6	10.0
Groundwater	13	0.4	88	1.9	1.1	587	0.7	1549	1.0	0.8
Grand total	2988	99.8	4619	100.0	99.7	86326	99.9	152196	100.1	100.0

are given, it should be noted that the distribution of contributions for 1982 is much more typical of the median or average situation under present land use conditions than is the distribution for 1981. There are several reasons for this. First, the runoff for 1981 was exceptionally low. The sum of flows for the 22 years of record at the four USGS gauges averages 185,600 acre-ft per year. The gauged flow for 1981 was 116,500, or 63% of the average. Although the runoff for 1982 was above the average (210,900), it was only slightly so (14%) and thus represents more accurately the median distribution of various contributions to the total nutrient loading of the lake. In addition, the wastewater treatment plant at Copper Mountain was operating under secondary treatment during 1981, whereas the standard of operation in the future, and for most of 1982, is tertiary treatment. Finally, the performance of the largest wastewater plant (Breckenridge) in 1981 was unusually poor due to equipment shutdown and malfunction.

The nutrient sources have been grouped in a number of ways in Table 60. In 1982, the more typical of the two years, about 16% of the phosphorus and 13% of the nitrogen came from the four major wastewater treatment plants combined. Package plants made only a minor addition to this total. Thus over three quarters of phosphorus and nitrogen loading under the present land use conditions comes from nonpoint sources of natural and man-made origin.

Climax Molybdenum contributed about 5% of the total phosphorus loading in 1981 and about half as much in 1982. The decline is explained by the reduced pace of operations, followed by shutdown in 1982. The nitrogen yield of Climax Molybdenum was greater than that of all four wastewater treatment plants combined.

Among the dispersed nonpoint sources, the background contribution, including everything that would be expected in the absence of human habitation, is con-

Table 60. Percentage Contributions to Loading of the Lake Aggregated in Three Different Ways[a]

Categories	P load (%)			N load (%)		
	1981	1982	Mean	1981	1982	Mean
Natural vs. man-made						
Natural	47	54	50	36	50	43
Man-made	53	46	50	64	50	57
Point vs. nonpoint						
Wastewater plants	38	16	27	20	13	16
Package plants and climax	7	3	5	33	16	24
All other sources	55	81	68	47	71	60
Human waste (WWTP's, package plants and septic) vs. all others						
Human waste	45	26	36	28	28	28
Other sources	55	74	64	72	72	72

[a] The 1982 data are most representative of a median year.

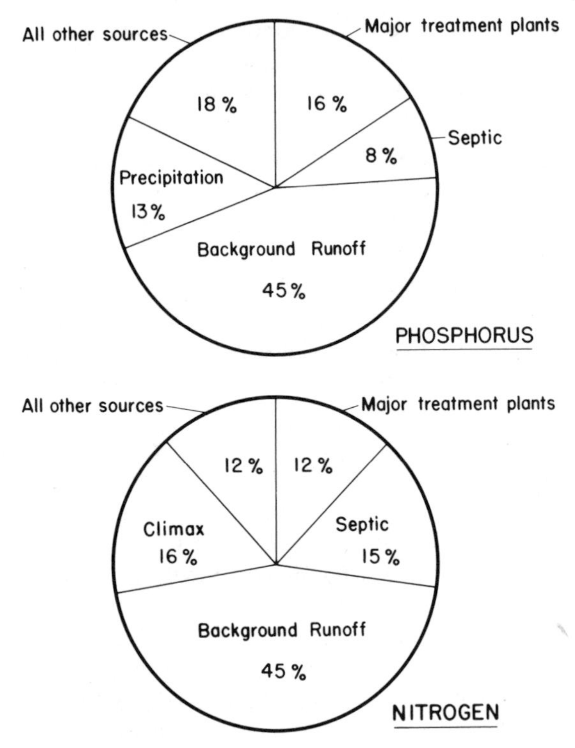

Figure 61. Percentage of loading due to various sources in 1982. All sources over 5% are shown separately.

siderably greater than all others combined. Under present conditions, it accounts for about half the total loading of nitrogen and phosphorus. Among the dispersed non-point sources related to human activity, septic systems stand out clearly as the highest contributor. All other sources account for a fraction of a percent to a few percent each. Although no single dispersed source related to human activity is of outstanding importance except for septic systems, the aggregate effect of the dispersed sources is significant.

Table 60 shows that the natural background and the sources associated with human presence share the phosphorus loading almost equally for phosphorus, and for nitrogen are only slightly skewed toward the man-made sources. Under the present land use conditions, in a year of median runoff, both the phosphorus and nitrogen loading of the lake will be divided approximately into four quarters. The first two of these quarters are taken up by the natural sources. The third quarter is taken up by sources associated directly with human waste (wastewater treatment plants, package plants, septic systems), and the fourth quarter is accounted for by human activity not related to the processing of human waste. In this last category are a wide assortment of mechanisms including earth disturbance, change in vegetative cover, fertilization, and others.

Figure 61 gives a visual impression of the nutrient contributions from various sources. The 1982 data are shown in the figure because these are most representative of the median of a run of years.

16. Modelling

One major purpose of the Lake Dillon study was to develop a model that could assimilate information on changes in land use and from this predict changes in the lake, including changes of aesthetic or economic importance. Such a model was developed from the information on the lake and watershed nutrient yields. The Lake Dillon Model consists of three parts, which are shown in diagrammatic form in Figure 62: a land use component, a trophic status component, and an effects component.

The Land Use Component of the Model

The land use component requires as input a matrix of land use information. The land use matrix specifies the intensity of different types of use on an area or population basis for each of 19 watershed segments (Figure 63). The land use component also requires the amount of gauged runoff (sum of the four USGS gauges) for the year that is being modelled, the amount of water to be pumped or diverted from other watersheds (a possibility for the future), and the mean concentration of this pumped or diverted water on a month-by-month basis. The land use component also accepts as input the water yield from all watershed segments, but will compute these water yields from the total gauged runoff if the water yields are not supplied. The nutrient export from each of the 19 watershed segments is computed from the water yield and the land use matrix by use of the equations that were developed in Chapter 14. The yields are summed for all segments and the inventory change function from Chapter 15 is

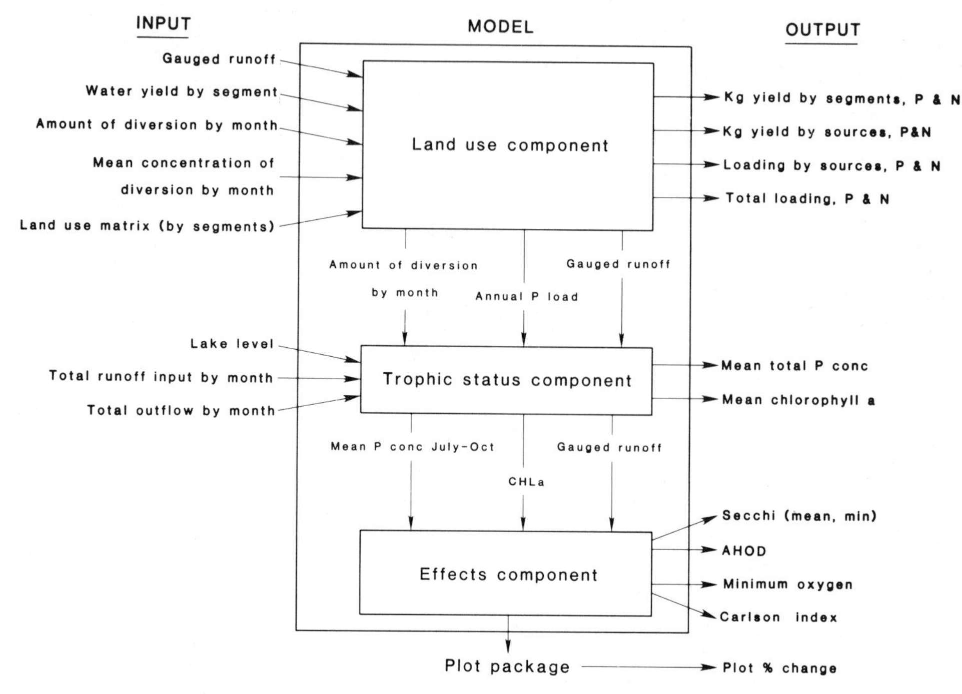

Figure 62. Structure of the Lake Dillon Model.

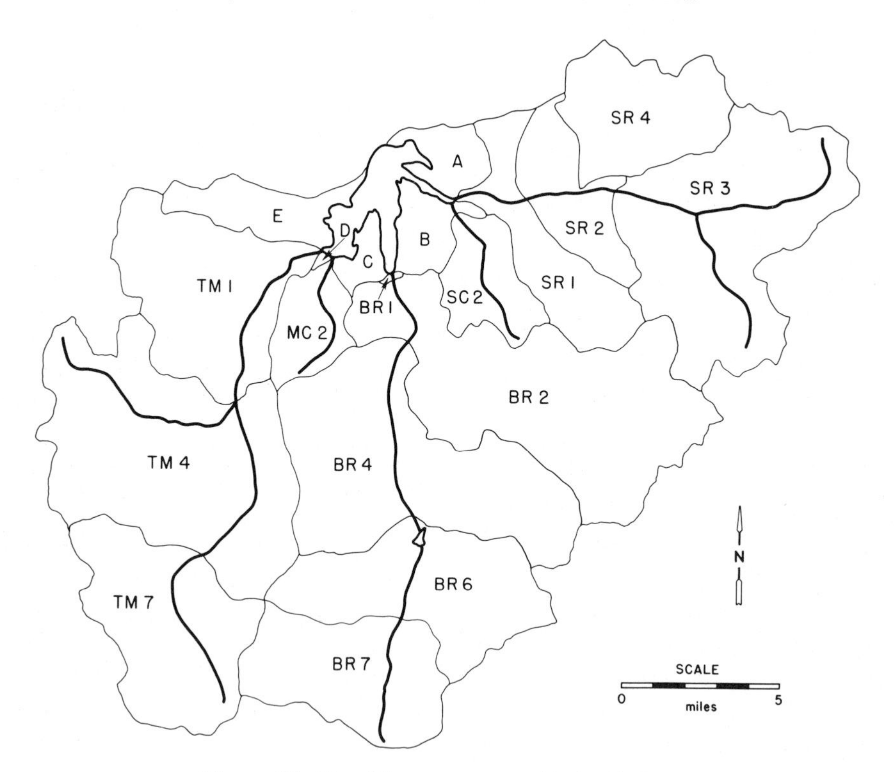

Figure 63. The 19 segments used in modelling.

applied. Other sources are then added, including the three lakeside WWTPs, groundwater, precipitation, diversions, and irregular construction activities. Output from the land use component of the model includes the nutrient yield by watershed segments and by source categories for both P and N and the total annual P and N reaching the lake.

The Trophic-Status Component of the Model

The land use component of the Lake Dillon Model passes certain information to the trophic-status component of the model. This information includes the monthly schedule of diversions or pumping, the concentrations of phosphorus and nitrogen on a monthly basis for pumped or diverted water, the annual phosphorus load by runoff and precipitation, and the amount of gauged runoff (i.e., sum of the four USGS gauges). Additional input required at this point but not supplied by the land use component includes the total surface water and ground water entering the lake on a monthly basis (this will be slightly larger than the gauged runoff), the monthly average lake level, and the amount of water leaving the lake on a monthly basis. The output includes the mean total phosphorus concentration of the top 15 m during the period July through October and the mean chlorophyll *a* in the top 5 m over the same interval.

Because the trophic status of lakes has been the subject of much modelling effort, there is considerable latitude in the choice of approaches for the trophic-status component of the model. Whereas with the land use component there is little basis for anything other than a simple empirical approach in modelling, the trophic-status component could be based in a more complex and fundamental way on current knowledge of biological, physical, and chemical processes in lakes. In discussing the alternative approaches, we recognize two different classes of models: process models (also called mechanistic models: Chapra and Reckhow 1983) and empirical models.

A process model assumes that the critical biological, chemical, and physical processes that have an important bearing on the variable of interest (in this case, trophic status and its derivatives) are sufficiently well known to be represented by equations. For example, a process model dealing with algal growth might include equations on nutrient uptake, redistribution of nutrients by sedimentation of algae through the metalimnion, and release nutrients to the water by algal decomposition. Since even the simplest natural planktonic systems are exceedingly complex at the elementary functional level, any such approach inevitably incorporates very large systems of coupled equations. An example is the Lake George Model, CLEANER (Park et al. 1979).

Process models have value as exploratory tools for basic ecosystem research, but their use for prediction, and thus for problem solving, may be ill-advised at present. Opinions are by no means undivided on this subject, but a consensus seems to be developing about the practical restrictions on process models (e.g., Talling 1979, Harris 1980, Hobbie and Tiwari 1977). The main reasons for the limitations on usefulness of process models are that the critical phenomena controlling biological variables in ecosystems are too poorly understood, too labile, and too complex to be treated in a reductionist manner. It is sometimes stated or assumed that the use of process models on ecosystems only awaits the accumulation of further basic information. The implication is that ecology is in a relatively primitive state and, as it matures, will logically support complex models that produce satisfying exact answers to practical problems. This line of reasoning is probably misleading in several ways. First, the fund of information presently available is quite vast, especially for lakes. It seems unlikely that any short-term improvement in this fund of information will drastically alter the feasibility of process models. Secondly, other disciplines in which models have long been used to deal with large and complex systems have encountered much the same limitations. Economics and meteorology are two such disciplines. The motivations for predicting either the economy or the weather well in advance are obviously enormous, yet the success of models to accomplish this is very modest; available models deal best with short-range predictions that assume nothing unusual happens.

There exists in the study of ecosystems, and particularly of lakes, an alternative to the process model. Lakes are so numerous that empirical experience can be accumulated on their responses to widely varying conditions and insults. This experience can serve as the basis for models. The philosophy of such models is entirely different from that of process models. The dependent variable of interest is examined from the viewpoint of a very small number of master variables. While these master variables are chosen for their known direct and indirect relationships to the dependent variable, no attempt is made to dissect the numerous separate functions that contribute to the overall significance of such a master variable. This approach was first popularized for lakes by Vollenweider (1968), who related trophic status to phosphorus concentration and phosphorus concentration to phosphorus loading and mean depth.

Empirical models inevitably evolve, taking on more complexity. While the general validity of such a model can be established by sampling large numbers of systems and comparing the observed to the predicted values for the variables of interest, there will always be a certain amount of scatter in the agreement of observations and predictions. An obvious challenge is to reduce the scatter by introduction of new concepts involving the same master variables or by addition of other master variables. The model thus tends to become more complex, and in principle converges with the process model by building backward from the level of master variables to the level of processes. It is unlikely that this convergence will ever be completed, however, as there are severe practical limits to the reduction of scatter in the relationship between observed and predicted values. Even in their most complex forms, present modifications of the Vollenweider model are very distant in rationale and complexity from a process model such as that developed for Lake George.

For reasons outlined above, we have accepted the empirical approach as a basis for the modelling of Lake Dillon. We have used a Vollenweider type of model, but have considered modifications as appropriate for the particular circumstances of Lake Dillon.

The first and simplest improvement in the original Vollenweider model, which was based solely on phosphorus loading and mean depth, involved corrections for the flow of water through lakes. Obviously two lakes with equal phosphorus loadings and mean depths but very different flushing rates might easily differ in trophic status. It seems intuitively obvious that a certain amount of the phosphorus income should be discounted for losses through the outlet. This discounting concept has been considered by a large number of students of eutrophication (Welch 1980, Chapra and Reckhow 1983).

It is a short step from corrections based on flushing rate to corrections based on empirically estimated phosphorus retention coefficients. This approach has been taken by Larsen and Mercier (1976) and by Dillon and Rigler (1974). Since phosphorus retention and flushing rate interact, a number of formulations are possible, as described by Vollenweider et al. (1980). Underlying all of these approaches is a general mass-balance equation by which a steady-state phosphorus concentration is defined in terms of phosphorus gains and phosphorus losses. In a lake such as Dillon, where flow-through is substantial, there is little doubt of the need to incorporate some kind of correction for the hydraulic residence time, or the sedimentation coefficient for phosphorus, or both of these.

The approach to be taken here can be credited to Vollenweider (1969), although it is a specific derivative of general mass-balance equations. Since this formulation was proposed by Vollenweider subsequent to his well-known original formulation, which did not take into account hydraulic residence time, we refer to it as the "modified" Vollenweider model. The equation is as follows:

$$C_{\mathrm{p}} = \frac{L_{\mathrm{p}}}{\bar{z}(1/t_{\mathrm{w}} + s)} \tag{16-1}$$

where C_{p} equals the phosphorus concentration of the lake, L_{p} is phosphorus loading of the lake, $\bar{z}$ is the mean lake depth, s is the phosphorus sedimentation coefficient and t_{w} is the residence time for water (Vollenweider et al. 1980). Vollenweider was able to reduce the number of parameters in the equation by assuming a sedimentation rate of 10–20 m per year. The simplified equation thus became

$$C_{\mathrm{p}} = \frac{L_{\mathrm{p}} t_{\mathrm{w}}}{\bar{z}(1 + t_{\mathrm{w}}^{0.5})} \tag{16-2}$$

The exact meaning of the model parameters is to some degree open to definition by the user. For Lake Dillon, we defined C_{p} as the mean phosphorus concentration of the upper 15 m of the water column over the interval between 1 July and 30 October. The layer 0–15 m corresponds to the maximum extent of the growth zone. Since changes in transparency and chlorophyll *a* in the upper water column are ultimately of interest, this was considered to be the most advisable choice for the definition of C_{p}. The time interval for determination of C_{p} was selected in such a way that the spring runoff, which incorporates an exceptionally large fraction of heavy particulate material that settles out rapidly, would not unduly influence C_{p}. This prevents the model from being unacceptably sensitive to variations between years in the amount of runoff.

The hydraulic residence time, t_{w}, can also be defined in a number of ways. In a lake that remains at steady state, all definitions are essentially equivalent. In the case of a lake such as Dillon, outflow and inflow are not always equal for a given year. For present purposes we use as our definition of t_{w} the ratio of lake volume to total inflow.

Inherent in the use of the Vollenweider equations to predict the phosphorus for individual years is the assumption that a new steady-state concentration can be approached each year. This assumption would not be reasonable for many lakes but is for Dillon because the hydraulic residence time is of the order of 1 year and the lake flushes rapidly in spring, thus setting the new phosphorus concentrations for the growing season to come.

The simplified formula (16-2) for the modified Vollenweider model predicts phosphorus values that are consistently too high for the 2 years of record. Since overestimates occurred both in a low-water and a high-water year, the implication is that the simplified formula underestimates the sedimentation coefficients for Dillon, and that the performance of the model could be improved by use of a more correct sedimentation coefficient, which would necessitate use of the more complex version of the relationship (equation 16-1). For this reason, sedimentation velocities (ν, where $\nu = \bar{z} \cdot s$) were approximated according to the procedure outlined by Higgins and Kim (1981). This procedure relies on knowledge of the inflow and outflow concentrations. The resulting estimates of sedimentation velocities for the 2 years were between 13 and 15 m/year. These sedimentation velocities were used to compute values of s ($\bar{z}$ was 23 m both years), which were in turn used in the Vollenweider equation (16-1). The predictions were then better but still consistently somewhat high for both years, implying that the actual sedimentation coefficient was still being underestimated. Since the estimation procedure from Higgins and Kim assumes a steady-state condition, which is not ever literally applicable, the estimate is subject to unknown error. Furthermore, the outflow concentrations could not be determined directly for that portion of the outflow water going through the Roberts Tunnel, which was not available for sampling, and this may have resulted in estimation error.

Since the sedimentation coefficients were underestimated even after correction by the procedure outlined by Higgins and Kim, another approach was taken. Each year was treated independently and the equation was solved for s using the observed values of the other parameters for the year in question. The resulting estimates of sedimentation velocity were 24 m/year for 1981 and 32 m/year for 1982. The larger coefficient for 1982 is quite reasonable in view of the fact that heavy particulates are moved in much larger quantity in years of high runoff than in years of low runoff (see Figure 22 in Chapter 5). The question then arises how the sedimentation coefficients should be set for purposes of prediction in years of different runoff. The positive relationship between sedimentation coefficient and amount of runoff seems justified from the information at hand but the exact form of the relationship cannot be determined from two points. It is therefore assumed that the relationship is linear. Nonlinearity will result in a certain amount of error, but the assumption of linearity is much better than assuming constancy of sedimentation, which is clearly incorrect.

A second element of the trophic-status component of the model is the relationship between phosphorus and chlorophyll. A number of statistically derived relationships are presented in the literature. These show consistently a statistically significant trend toward higher summer chlorophyll *a* with higher total phosphorus in the mixed layer. There is, however, quite a bit of scatter and this has been the subject of considerable discussion and analysis.

The general equations that are available in the literature (e.g., Dillon and Rigler 1974, Carlson 1977, Jones and Bachmann 1976, Jones and Lee 1982, Lambou et al. 1982) do a uniformly poor job of predicting Lake Dillon chlorophyll *a*. Observations are three to six times higher than predicted values; Dillon produces considerably more chlorophyll per unit of P than most lakes. This is unusual but not unique, as shown by the scatter of points in the above-cited publications, but the explanation is not immediately obvious.

A recent study by Smith (1982) appears to explain the unusual phosphorus-chlorophyll relationship of Lake Dillon. Smith assembled data from several surveys in which both the nitrogen and phosphorus concentrations were known. He treated nitrogen as an added variable in studying the relationship between phosphorus and chlorophyll, and showed that nitrogen is a very important covariate. Use of nitrogen as well as phosphorus in regression equations greatly increases the amount of variance that can be explained even when phosphorus clearly limits growth rates. His conclusion is that increases in the nitrogen to phosphorus ratio boost the amount of chlorophyll that will be produced for a given amount of phosphorus, even when phosphorus is limiting. The mechanism by which this occurs is unknown, but the reduction of variance by incorporation of nitrogen into the equations is impressive. It is worth noting that OECD (1982) found no added variance component explained by nitrogen. Smith and OECD did use different data bases; possibly Smith's was better suited for statistical illustration of the role of nitrogen. For present purposes, we accept Smith's conclusions as valid and attempt to apply them to Lake Dillon.

The lakes studied by Smith varied widely in the ratio of nitrogen to phosphorus. Smith developed equations for predicting chlorophyll in lakes of differing nitrogen to phosphorus ratios. His equations show that the effect of increasing the nitrogen to phosphorus ratio is to shift the line relating phosphorus to chlorophyll upward (Figure 64). At low to moderate TN:TP ratios, the upward shift occurs without changes in slope; the relationship thus takes the form of a family of parallel lines. In the highest category used by Smith, however (TN/TP > 35 by weight), the slope increases, and simultaneously becomes insensitive to further change in the TN:TP ratio. For lakes with low TN:TP ratios, it is essential to use N as an independent variable as well as P. For lakes with high TN:TP ratio, N can be dropped from the equation, but the coefficient on P will differ from that of lakes having low TN:TP ratios.

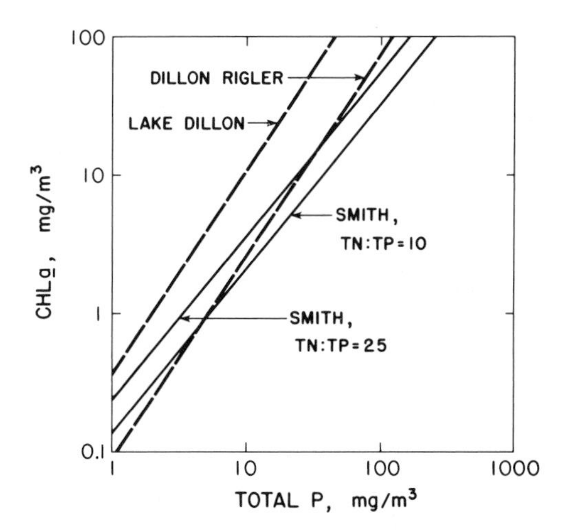

Figure 64. Log-log plot of total P and chlorophyll *a*, showing lines for two of Smith's (1982) lower TN:TP categories, the Dillon-Rigler line, and the Lake Dillon line.

Lake Dillon, with TN:TP = 54 (weight basis), fails in the very highest category of nitrogen to phosphorus ratios among the lakes studied by Smith. In this highest category, as the influence of further change in TN:TP becomes negligible, the coefficient on P is high and convergent with that of the Dillon–Rigler equation (1974). This convergence is expected, since the Dillon–Rigler equation was developed without consideration of N. However, Smith's study shows that the intercept may continue to be affected as TN:TP is raised to high levels, even though the slope stabilizes. We thus approach the Dillon data with the following points in mind: (1) the TN:TP ratio is so high in Dillon that N need not be used in the equation, (2) the appropriate slope can be taken from the Dillon–Rigler equation, (3) the intercept should be determined uniquely for Dillon. We therefore set the slope of the phosphorus–chlorophyll equation equal to that of the Dillon–Rigler equation, but assume that the intercept will be determined by the nitrogen to phosphorus ratio. We determined the intercept empirically by solving for the intercept using the slope from the Dillon–Rigler equation and the observed chlorophyll *a* values. The resulting equation is as follows:

$$\log B = 1.449 \log(C_{\mathrm{p}}) - 0.398 \qquad (16\text{-}3)$$

where B is chlorophyll *a* (July–October, μg/liter, 0–5 m) and C_{p} is total P (0–15 m, July–October, μg/liter). This equation performs well for both years (predictions will be given below), suggesting that the Dillon–Rigler slope is in fact appropriate for Lake Dillon.

Effects Component of the Model

Total phosphorus and chlorophyll *a,* as predicted by the trophic status component of the Lake Dillon Model, have a number of correlates of economic and aesthetic importance. Since public attention and the attention of lake managers and users is likely to be focussed on some of these correlates, it is useful to be able to make predictions for them.

Transparency is often of greatest concern as a correlate of eutrophication. The most common measure of transparency is the secchi depth. Since a great deal of information is available from the literature on the relationship between secchi depth and chlorophyll *a,* and since there is a complete set of secchi depth measurements for Lake Dillon in 1981–1982, we have designed the effects component of the model to predict secchi depth. The prediction is based on chlorophyll *a,* which is the variable most directly responsible for changes in transparency as eutrophication occurs. The interval of interest is 1 July through October, corresponding with the postrunoff stratification season. As in other instances, we have excluded the month of June from the predictions because all predictions are complicated in this month by the high amount of inorganic particulate matter in the water column as a result of runoff. The observed mean secchi depth over the July–October interval in 1981 was 3.15 m and in 1982 it was 2.76 m.

The literature offers a number of equations relating secchi depth to chlorophyll *a.* There are two drawbacks to the use of these general equations on Lake Dillon. First, there is a great deal of scatter around any of the lines taken from the literature, as would be expected for the comparison of very different kinds of lakes. Second, Lake Dillon is not located in the densest cluster of points used for any of the equations, since its chlorophyll *a* values are relatively low. Thus the general equations, which for transparency are particularly likely to go astray when extrapolated from lakes of low transparency to those of high transparency, may not be especially appropriate for Lake Dillon.

The relationship between secchi depth and chlorophyll *a* is always nonlinear; transparency as measured by secchi depth decreases at a slower rate than chlorophyll increases. The fit of points to an empirical transparency/chlorophyll *a* relationship is typically rendered linear by log transformation. Problems related to the determination of the slope and the intercept of this relationship are different. The slope, which shows the increment of change in secchi depth for an increment of change in chlorophyll *a*, is much more universal than the intercept. The package size of chlorophyll *a* (i.e., the average cell size for phytoplankton) results in different values of chlorophyll-specific extinction (Kirk 1975, Harris 1978), and this may be responsible for a certain amount of variation in the slope. We have already shown, however, that Lake Dillon is very average with respect to its chlorophyll-specific extinction (e_s = 0.015 m^2/mg chlorophyll *a*). On this basis it seems justified to use a slope from one of the equations in the literature. We choose for this purpose the slope of the equation of Carlson (0.68), which is based on 147 lakes and has an r value of 0.93 (Carlson 1977). The National Eutrophication Survey data produced a higher slope (0.86) (Lambou et al. 1982), but the observed values for Lake Dillon over the two years of record suggested much closer agreement with the Carlson coefficient than with the National Eutrophication Survey coefficient.

The intercept of the log-transformed relationship between secchi depth and chlorophyll *a* is likely to vary much more extensively between lakes than the slope, particularly if one considers reservoirs and natural lakes together. The intercept for a given lake will be determined largely by a combination of nonchlorophyll factors contributing to the extinction of light. In very transparent lakes, this includes principally the extinction effect of pure water and of dissolved substances. In lakes that receive substantial amounts of inorganic particulates, extinction by particulates becomes a major determinant of the intercept.

It is feasible to adapt the intercept for a prediction equation to a given lake by using a general slope such as that of Carlson (1977) along with the observed values for the lake in question to solve for the intercept. We did this separately for 1981 and 1982 on the Lake Dillon data using the Carlson slope. The intercept for 1981 was 2.44 and the intercept for 1982 was 2.37. These intercepts were so similar that a common intercept of 2.4 was adopted for use with the Dillon data. This intercept is higher than the intercept of the Carlson equation (2.04), probably because the Carlson equation is based on natural lakes lacking the inorganic particulate load that one sees in Lake Dillon. The intercept is much closer to but slightly lower than the intercept for the National Eutrophication Survey (2.56), which includes a large number of reservoirs.

The equation resulting from the above analyses is as follows:

$$\ln \mathrm{SD} = 2.4 - 0.68 \ln B \tag{16-4}$$

where SD is secchi depth in meters and B is chlorophyll *a* in micrograms per liter.

A rough approximation can also be made of the minimum secchi depth from the average secchi depth. The minimum secchi depth in this case applies to the time of maximum phytoplankton standing stock. There is another seasonal minimum associated with runoff, which is not of concern here. In 1981 the ratio of minimum secchi depth to average secchi depth was 0.76 and in 1982 it was 0.65. As an approximation we take the ratio to be 0.70. There is no extensive literature on this subject and the prediction is therefore less secure than the other predictions mentioned up to this point. However, the use of this ratio will give some indication of the transparency of the lake at that particular time of the year when transparency problems may be most aggravated (July).

Oxygen depletion in deep water during the summer is also of concern in connection with eutrophication. In the section on oxygen, we presented an analysis of the areal hypolimnetic oxygen deficit and showed good agreement between the predicted AHOD and an equation developed by Cornett and Rigler (1980) based on secchi depth and mean lake depth. We use this equation in the effects component of the Lake Dillon Model, but we believe it is feasible to improve the prediction by making a correction for the amount of water entering the lake in a given year. In 1981 the equation underpredicted the AHOD by 130 mg/m^2/day (20%) and in 1982 it overpredicted by 6 mg/m^2/day (1%). The difference in agreement of observed and expected in the 2 years is probably due either to high metalimnetic production in low water years or to higher amounts of oxygen transported by advection to deep water in wet years. An empirical correction is possible by making the assumption that runoff is proportional to the differential between the predicted and actual hypolimnetic oxygen deficits.

Although the AHOD is of some direct interest, the minimum oxygen level in deep water is of more specific concern. There are no general equations for prediction of minimum oxygen content, since this prediction would involve the shape of the lake and the duration of the stratification period. However, for a given lake, there is obviously a very close relationship between the AHOD and the amount of oxygen depletion. The relationship will be very close to linear if the initial oxygen and the duration of the stratification period are constant. For Lake Dillon, the initial oxygen can safely be assumed equal to 9.0 mg/liter, the saturation concentration at spring overturn, and the stratification will be close to constant duration. The oxygen depletion is equal to the difference between 9.0 mg/liter and the minimum oxygen observed toward the end of stratification. In developing an equation to make predictions, we deal specificially with a point 5 m over the bottom of the index station, and use the degree of depletion (mg/liter) in each of the 2 years as the dependent variable and the AHOD (mg/m^2/day) as the independent variable. This yields a slope of 0.00698. This slope, when multiplied times the AHOD, gives the predicted amount of depletion (mg/liter). The amount of oxygen remaining is then the difference between the depletion and 9.0. The lower bound of remaining oxygen is set to zero for depletion more than severe enough to remove all oxygen.

Comparing Predictions and Observations for 1981 and 1982

The complete Lake Dillon Model was run on land use and runoff data for 1981 and 1982. In these runs, the unusual features of 1981 or 1982 that would not be obvious

Table 61. Comparison of Predictions from the Lake Dillon Model with Observations for 1981 and 1982

	1981		1982	
	Predicted	Observed	Predicted	Observed
Total P (μg/l)[a]	6.7	7.0	7.1	7.4
Chlorophyll *a* (μg/l)[a]	6.3	6.7	6.9	7.3
Mean secchi (m)[a]	3.2	3.2	3.0	2.8
Minimum secchi (m)	2.2	2.4	2.1	1.8
AHOD (mg/m^2/day)	706	710	590	630
Minimum deep-water O_2	4.1	4.4	4.9	4.6

[a] Mean, July–October.

from runoff or from land use data of the type accepted by the model for prediction purposes were ignored. For example, the output of wastewater treatment plants was computed from the number of persons served using the general relationships described in Chapter 14 (Table 57) rather than the known yields from these plants. Similarly, the construction activity in the Snake River bottom near Keystone in 1982 was not treated in any special way in the model for 1982. Thus the agreement between predictions and observations for the two model runs gives an idea of the deviation between observed and expected that could be produced by randomly occurring events whose timing cannot be anticipated.

The predictions and observations are summarized in Table 61. For all variables the agreement is within 5%, except for the minimum deep-water oxygen, which is within 10%. The performance of the model on these years of known characteristics is thus excellent, despite the occurrence of a number of events whose timing cannot be anticipated by such a model.

17. Using the Model for Prediction

The Lake Dillon Model was applied to eight different scenarios considered to encompass the range of possibilities that might be realized over the next 20–25 years. The scenarios comprise five sets of assumptions for development and land use, each of which was applied to hydrologic assumptions for a dry year and for a wet year. Predictions in each case include the total loading of the lake, the source distribution of the loading, and the response of the lake to the predicted amount of loading. The dry-year hydrologic conditions were always set identical to 1981, and the wet-year conditions were set identical to 1982. Use of the hydrologic conditions of 1981 and 1982 is advantageous since the behavior of the lake in these 2 years under present land use is well documented and can thus be compared easily to future years of similar hydrologic conditions but very different land uses.

The model predicts trophic indicators and these are translated into a trophic-status category. This requires certain conventions. We use chlorophyll as the key trophic indicator, since algal biomass is the central concern of changing trophic status. We set limits on the trophic categories from a midrange of boundaries accepted by various sources. According to the review by Welch (1980), the oligotrophic boundary will be somewhere between 2 and 4 μg/liter and the eutrophic boundary will be between 6–10 μg/liter. We therefore set the mesotrophic range for present purposes at 3–8 μg/liter chlorophyll *a*. This is also the range used by OECD (1982). Another convention is the adoption of a one-year interval as the minimum time over which a trophic classification can be resolved. Trophic classes make no sense at intervals less than one year, so one year is the maximum resolution possible. There is some merit to the argument that trophic status should be based on a run of years, or on a year of median conditions. However, since we are also interested in variation between years, we adopt for modelling purposes an annual interval.

Model Runs for Present Land Use Patterns

Before any of the scenario data were used, the model was run twice (wet year/dry year) with a land use input matrix representing present conditions. The purpose of these model runs was not to check the behavior of the model, which has already been discussed, but rather to define the loading and trophic condition of the lake under standardized conditions subsequent to 1982. These model runs differ from the predictions discussed previously in that Copper Mountain is assumed to have transferred completely to tertiary treatment, and to serve 2500 persons, slightly more than 1981–1982. Climax Molybdenum is set at the 1981 employment level (2650 persons). Also, no allowances are made for any of the special events that occurred in 1981 and 1982. Thus these two model runs for present conditions give for comparative purposes the output of the model under the assumption that present land use stabilizes (by resumption of activities at Climax and completion of the transfer of Copper Mountain to tertiary) but that no growth occurs. The results of the two model runs on stabilized present conditions are shown in Tables 62 and 63. As expected, the predicting loading and response of the lake are very similar to those actually observed in 1981 and 1982.

Scenarios 1A, 1B: Low Growth

The four wastewater treatment plants were meeting their wasteload allocations in 1981–1982. Thus one possibility is that most additional growth would involve waste disposal through septic systems. Scenarios 1A and 1B assume that the number of persons on septic systems doubles, but that treatment plant discharges remain as they are. All other land uses remain unchanged.

The model was run first for a wet year (1A) and then for a dry year (1B). The results are reported in Tables 62 and 63. Figure 65 gives a graphical representation of the degree of change as compared with present conditions. In the wet year, the predicted loading is higher by slightly less than 10% as a result of low growth. This causes an increase of chlorophyll to 8 μg/liter. Since we have taken the range of chlorophyll values between 3 and 8 μg/liter as indicating a mesotrophic condition, the low-growth scenario pushes the chlorophyll *a* exactly to the upper limit of the mesotrophic span in a wet year. There is an accompanying decrease in transparency and a decrease in the minimum oxygen at the bottom of the lake.

For the dry year, the degree of change in total loading is smaller. Furthermore, the lake is much more toward the middle of the mesotrophic range for chlorophyll and other trophic indicators during a dry year, except for deep-water oxygen. Thus increasing the loading slightly to the degree specified by the low-growth scenarios does not bring the chlorophyll level so close to the upper boundary of the mesotrophic range. For a dry year, the lake remains solidly mesotrophic.

Table 62. Results of the Application of the Lake Dillon Model to Scenario Data

Scenario Conditions	Total P Load (kg)	Percentage WWTP[a]	Percentage Septic	Percentage Diversion	Percentage Background[b]	Percentage (Other)
Wet year (similar to 1982)						
Current conditions[c]	4678	17.4	8.2	0	60.5	13.9
Low growth (1A)	5049	16.2	14.7	0	56.1	13.0
High growth (3A)	6389	12.8	31.9	0	44.4	10.9
High growth, perfect controls (4A)	3657	22.3	0	0	77.5	0.2
High growth, partial controls (5A)	5001	16.3	20.4	0	56.6	6.7
Dry year (similar to 1981)						
Current conditions[c]	2459	30.0	6.6	0	51.7	11.7
Low growth (1B)	2618	28.3	12.0	0	48.5	11.2
High growth (3B)	3190	23.2	27.2	0	39.9	9.7
High growth, perfect controls (4B)	2021	36.6	0	0	62.9	0.5
High growth, partial controls (5B)	2596	28.5	16.7	0	49.0	5.8

[a] Includes package plants.
[b] Includes background runoff and precipitation.
[c] Assumes Copper Mountain on teritary, Climax mines at full employment 1981 levels.

Table 63. Results of the Application of the Lake Dillon Model to Scenario Data

Scenario Conditions	Lake Total P ($\mu g/l$)	Chlorophyll ($\mu g/l$)	Mean Secchi (m)	Minimum Secchi (m)	Minimum O_2 (mg/l)	Trophic Status
Wet year (similar to 1982)						
Current conditions[a]	7.3	7.1	2.9	2.0	4.8	Mesotrophic
Low growth (1A)	7.9	8.0	2.7	1.9	4.6	Mesotrophic
Low growth with diversions (2A)	57.0	141.0	0.38	0.27	0.0	Eutrophic
High growth (3A)	10.0	11.2	2.1	1.5	3.8	Eutrophic
High growth, perfect controls (4A)	5.7	5.0	3.7	2.6	5.4	Mesotrophic
High growth, partial controls (5A)	7.8	7.9	2.7	1.9	4.6	Mesotrophic
Dry year (similar to 1981)						
Current conditions[a]	6.1	5.4	3.5	2.4	4.3	Mesotrophic
Low growth (1B)	6.4	6.0	3.3	2.3	4.2	Mesotrophic
Low growth with diversions (2B)	74.0	206.0	0.29	0.21	0.0	Eutrophic
High growth (3B)	7.9	7.9	2.7	1.9	3.6	Mesotrophic
High growth, perfect controls (4B)	5.0	4.1	4.2	4.8	4.8	Mesotrophic
High growth, partial controls (5B)	6.4	5.9	3.3	3.0	4.2	Mesotrophic

[a] Assumes Copper Mountain on tertiary, Climax mines at full employment 1981 levels.

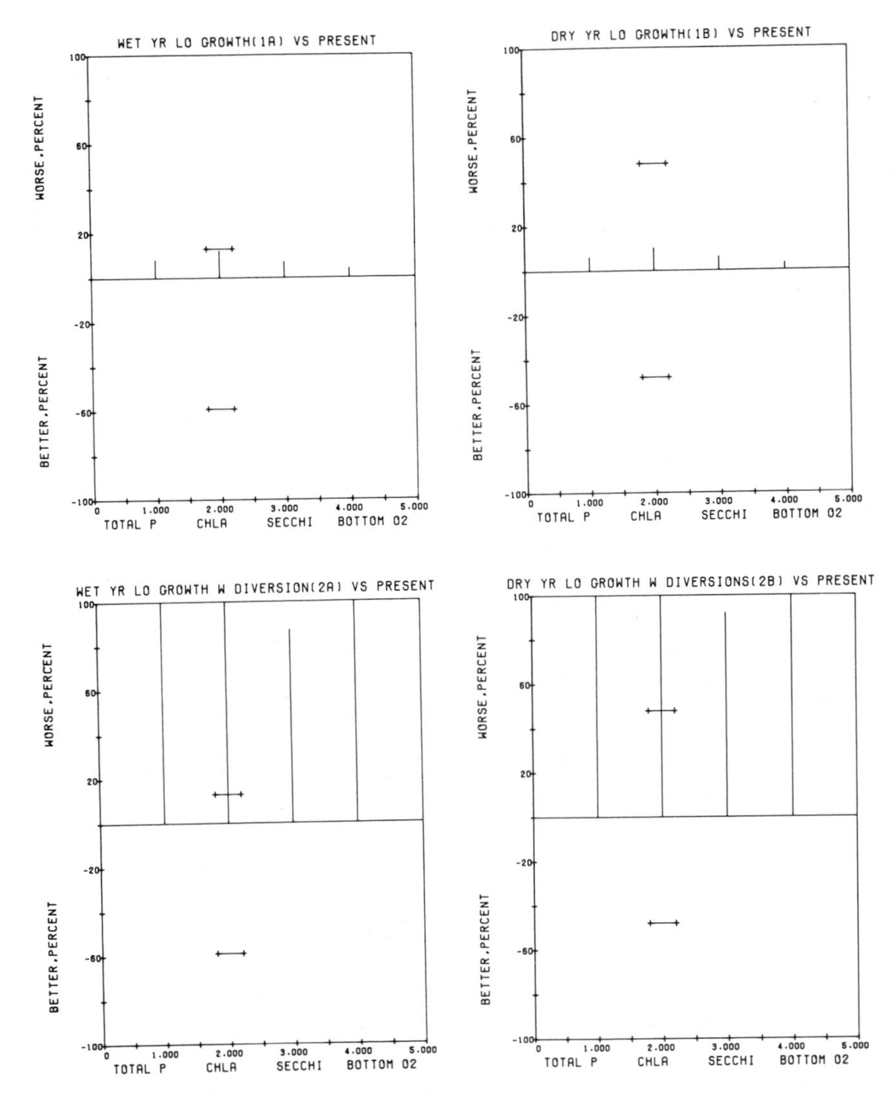

Figure 65. Graphical summary of model output for four key lake characteristics. Vertical lines show deviation from present conditions; horizontal lines with "+" mark the oligotrophic and eutrophic boundaries for chlorophyll *a*.

Scenarios 2A, 2B: Low Growth with Diversions

Since Lake Dillon is the water storage facility for Denver, whose water needs are growing, one possibility is augmentation of water flowthrough by diversion of additional Western Slope water to the lake. In scenarios 2A and 2B it is assumed that 10,000 acre-ft/year is diverted into Lake Dillon from Straight Creek (Figure 54) and 183,000 acre-ft/year is diverted into Dillon by the Eagle-Colorado Project, which would bring in water from the Gore Range. The total phosphorus concentrations are set at 30 μg/liter for Straight Creek and to 200 μg/liter for Eagle-Colorado. These concentrations were chosen as the worst case and obviously represent considerable enrichment of the diverted water by nutrient sources outside the watershed. The amounts of water diverted are assumed to apply to years of average or above-average moisture. For the dry-year run of the model (2B), the amounts of diversion are scaled down in proportion to the gauged runoff (sum of the four USGS gauges in the watershed). The diverted water is assumed to enter the lake on a schedule identical to the Tenmile Creek hydrograph. All other conditions are set identical to scenarios 1A and 1B.

The results of the two model runs (2A, 2B) are summarized in Tables 62 and 63 and Figure 65. The phosphorus loading coming in by way of diversion is roughly 10 times the loading under present conditions without diversions. The lake shows drastic increases in total phosphorus concentrations and in chlorophyll levels. Complete oxygen depletion is predicted for the hypolimnion, and predicted secchi depth values are extremely low. The lake under these circumstances would be unequivocally upper eutrophic. Nuisance blooms would be a certainty under these conditions.

The chlorophyll levels predicted for the diversion scenarios are so high that they probably exceed the theoretical maximum chlorophyll. About 300 mg/m^2 of chlorophyll *a* absorbs 99% of the light (Margalef 1978, Talling 1982). Light rather than nutrients thus becomes limiting. In the upper 5 m of Lake Dillon, to which algal growth would be essentially confined under low transparency conditions, we could expect to find at a maximum about 60 μg/liter of chlorophyll *a*, which would account for 300 mg/m^2 summed up over the 5-m mixed layer. Thus the chlorophyll predictions, which are in excess of 100 μg/liter, merely indicate that the lake reaches the maximum biomass possible under the light limitation conditions prevailing in Lake Dillon. The exact concentrations are of course of little interest once they reach such high levels.

Scenarios 3A, 3B: High Growth

These scenarios assume that current wasteload allocations are met for the wastewater treatment plants. The number of persons served by these plants is therefore allowed to remain as for scenarios 1A and 1B, and Climax Molybdenum is assumed to operate with a workforce of 2650 persons. The high-growth scenarios assume that septic systems will serve a peak population of 37,000 persons. This corresponds to a time-weighted average of 8409 full-time equivalent residents. It is assumed that this number of persons is distributed within the watershed according to the same pattern of present septic system users.

Tables 62 and 63 and Figure 66 summarize the results of the model runs for high growth in a wet year (3A) and in a dry year (3B). For the wet year, the scenarios indicate an increase in total wet-year loading of approximately 50% over present conditions. There is a drastic increase in the percentage of loading caused by septic systems, as would be expected. Chlorophyll is predicted at 11.2 μg/liter, which is well over the boundary from mesotrophic to eutrophic. Transparencies and minimum oxygen in deep water similarly reflect a higher trophic status. For the dry year, the effects of high growth are of smaller magnitude but still significant. Chlorophyll is predicted to fall at 7.9 μg/liter, just below the upper mesotrophic boundary at 8 μg/liter. Thus the

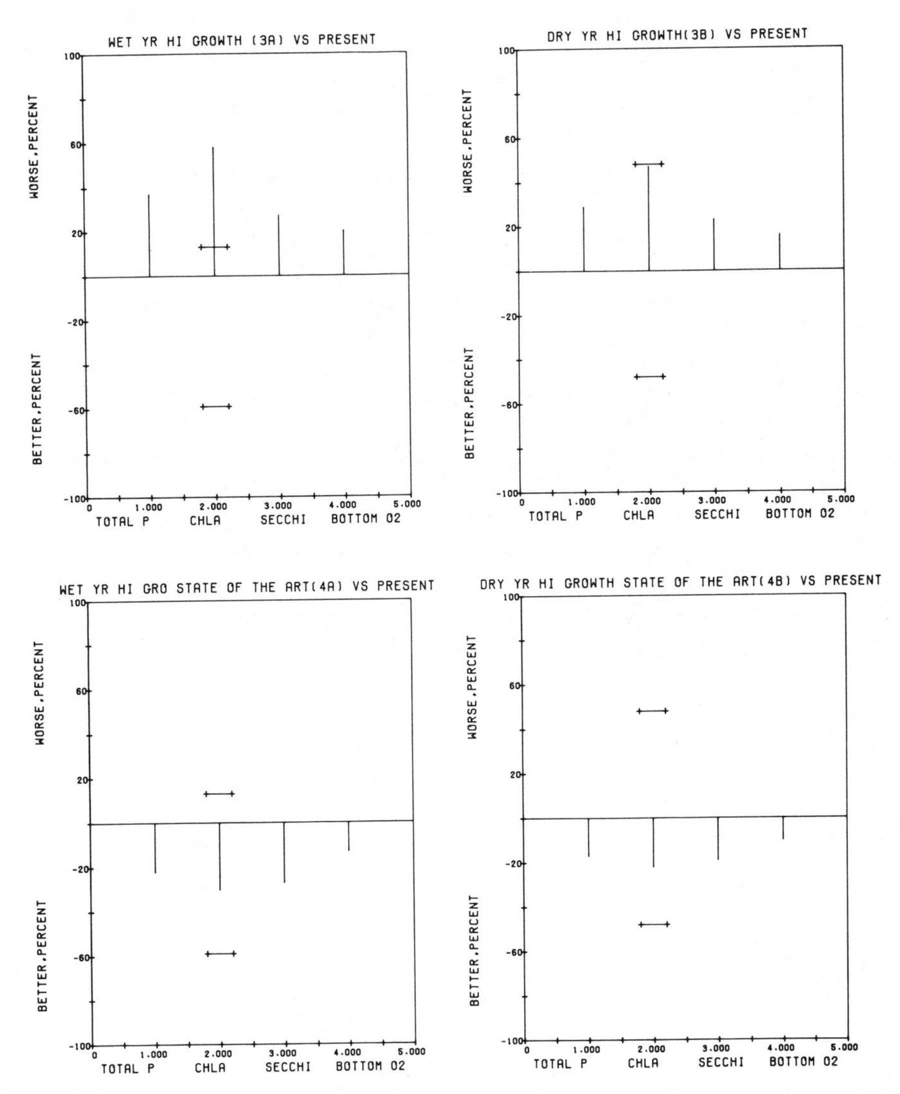

Figure 66. Graphical summary of model output for four key lake characteristics. Vertical lines show deviation from present conditions; horizontal lines with "+" mark the oligotrophic and eutrophic boundaries for chlorophyll *a*.

lake could probably be classified in the very dryest of years as mesotrophic, but at all other times would be eutrophic, and in wet years would be especially so.

Scenarios 4A, 4B: High Growth with Perfect Nonpoint Controls

Nonpoint source controls are being considered as a means of controlling trophic status while allowing population growth. Scenarios 4A and 4B specify that wasteload allocations are met by point sources, and that no diversions enter the lake. All sources except the wastewater treatment plants, package plants, groundwater, precipitation, and background are set to 0. Thus it is assumed that all nonpoint sources are completely eliminated.

The results of the model runs are shown in Tables 62 and 63 for a wet year (4A) and for a dry year (4B). Figure 66 shows the corresponding graphs. In either a wet or a dry year, the model predicts significant improvement of the trophic status of the lake, despite high growth. Chlorophyll values show lower mesotrophic status for the lake under this scenario. Transparency, bottom oxygen, and other indicators are correspondingly improved.

As attractive as this scenario seems, it is unrealistic. It implies that there would be no yield whatever above background for septic systems, for Climax Molybdenum, and for the many distributed sources of nutrients associated with human activity. The high-growth conditions imply that about 8000 persons (annual average) will be served by septic systems. This is essential because the wasteload allocations cannot be held constant if additional population is to be served on sewer. Thus a substantial septic contribution or a change in the wasteload allocation seems inevitable as an accompaniment to high growth, but scenarios 4A and 4B allow neither. The results should therefore be viewed as representing an unreachable upper bound of control over phosphorus loading, regardless of financial considerations.

Scenarios 5A, 5B: High Growth with Partially Effective Nonpoint Controls

Scenarios 5A and 5B are intended as counterparts to scenarios 4A and 4B, but with controls set at levels that are more likely to be realized. The assumptions are identical

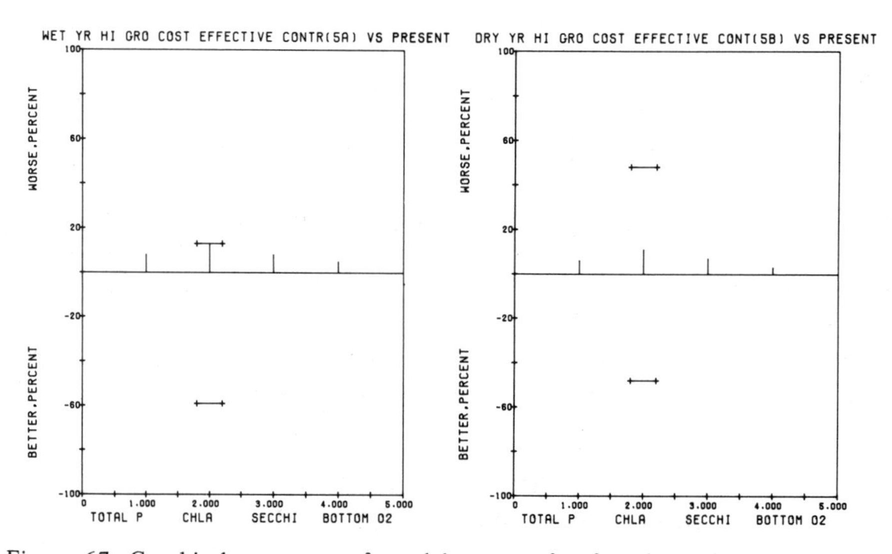

Figure 67. Graphical summary of model output for four key lake characteristics. Vertical lines show deviation from present conditions; horizontal lines with "+" mark the oligotrophic and eutrophic boundaries for chlorophyll *a*.

to those of scenario 4A and 4B, except that all nonpoint sources, which were set to 0 in 4A and 4B, are allowed to assume half their uncontrolled value. The results appear in Table 62 and 63 and Figure 67.

For a wet year, the model predicts loading that is slightly more than 5% above present conditions. This causes a slight increase in the expected chlorophyll level, which reaches 7.9, very close to the upper mesotrophic boundary at 8.0. For a dry year, there would also be a slight increase in chlorophyll above the present conditions, but the dry-year increase would not bring the chlorophyll so close to the upper mesotrophic boundary. Thus the model shows that, if the controls could be accomplished, the lake would in most years fall in the mesotrophic range. Although scenarios 5A and 5B may be technically realistic they may actually imply intolerable expenditures.

Overview of Scenario Studies

The modelling indicates that Lake Dillon will move into the eutrophic category if diversion water rich in phosphorus is added in quantity to the lake or if high growth occurs without the adoption of nonpoint source controls or other measures not now in practice to reduce the phosphorus loading of the lake. Under low growth or high growth with additional controls, the condition of the lake could be held within the mesotrophic range, but might suffer some trophic degradation or remain more or less the same, depending on the exact conditions.

18. Summary

1. The purposes of the Lake Dillon Study were as follows: (a) to provide comprehensive limnological information on the lake, including its seasonal cycles and its basic physical, chemical, and biological features with special attention to trophic status; (b) to provide information on the present nutrient sources of the lake and their relation to land use; and (c) to construct a model that would be capable of predicting the trophic status of the lake given any likely combination of future development patterns for the Lake Dillon catchment.

2. The limnological study showed that the seasonal events in Lake Dillon can be divided as follows: (a) ice cover from January through April, (b) spring mixing from early May to the end of May, (c) summer stratification from June through October, and (d) fall mixing for most of November and all of December. During summer stratification, the mixed layer is stable at a thickness of 5-10 m until September, when it begins to thicken.

3. Transparency is lowest during the first half of the stratification period. There is a sudden decrease in transparency in June caused by the entry of large amounts of particulate material from the watershed. Depending on the amount of runoff, this reduces the transparency at the surface from about 4 m to less than 1 m. A large proportion of the entering particulate material settles out quickly, but rapid growth of algae causes transparency to stay low. Minimum secchi depths in July, at the time of the chlorophyll maximum, are between 1.5 and 2.5 m. Transparency increases steadily after the chlorophyll maximum in July because of algal nutrient depletion. Fall transparencies are high (secchi, 3-5 m).

4. Runoff from the three rivers enters the lake at different depths depending on the wetness of the year (amount of runoff) and the time of year. During both 1981 and 1982, river water entered the lake at progressively greater depths during the course of the stratification season. Entry level was uniformly higher in 1982, when there was much runoff, than in 1981. In both years runoff entered the upper water column in the early period of stratification, but entered the middle water column during the middle and late portions of stratification.

5. Annual average total phosphorus over all depths in Lake Dillon was 6.6 μg/liter in 1981 and 7.7 μg/liter in 1982. Peak values at the surface during runoff were considerably higher than this in both years, however (12–17 μg/liter). Soluble inorganic phosphorus concentrations were consistently very low. The upper water column showed a steady loss of total phosphorus from the beginning of stratification until the increase in thickness of the mixed layer in September. Soluble organic phosphorus was almost totally depleted by the end of August.

6. Total nitrogen in the entire water column averaged 294 μg/liter in 1981 and 447 μg/liter in 1982. Lake Dillon is unusual in its high ratio of total nitrogen to total phosphorus. This is partly due to tertiary treatment, which removes phosphorus but not nitrogen, and to mining, which releases much nitrogen into upper Tenmile Creek. Soluble inorganic nitrogen is steadily depleted from the upper water column after the onset of stratification. Depletion of inorganic nitrogen is essentially complete by the middle of July.

7. The amount of phosphorus in the sediments of Lake Dillon does not differ from the amounts typical of natural oligotrophic lakes.

8. Total particulates in the upper water column (0–5 m) averaged 2.3 mg/liter in 1981 and 3.8 mg/liter in 1982. In both years there were two peaks of total particulates in the upper water column: one at the time of runoff caused by inorganic particulates, and a second in July caused by phytoplankton biomass.

9. The maximum chlorophyll *a* at the surface occurred during July of both years and fell between 11 and 13 μg/liter. The average for the postrunoff stratification season was 6.7 μg/liter in 1981 and 7.3 μg/liter in 1982. The highest chlorophyll values at any depth in both years were between 17 and 18 μg/liter.

10. Repeated elements of the annual composition cycle for phytoplankton were *Asterionella,* which is abundant under the ice and during early spring in Lake Dillon, and *Synedra,* which accounts for a major portion of the July peak of chlorophyll in Lake Dillon. Irregular but very large populations were observed of the blue-green alga *Synechococcus* and the diatom *Rhizosolenia.* No nitrogen-fixing blue-greens were found. Zooplankton composition was simple, consisting almost entirely of three rotifer species and one copepod species. Calculation of grazing rates shows that zooplankton are a minor source of phytoplankton mortality most of the year but may be the major source of mortality in September.

11. Low ratios of phosphorus, nitrogen, and chlorophyll *a* to organic matter during stratification suggest significant nutrient stress on the phytoplankton during stratification. The ratio of maximum photosynthesis to chlorophyll *a* was highest during periods of deep mixing and lowest under ice cover and during the first half of stratification, indicating that the greatest nutrient stress occurs under ice and between the middle of June and the end of August. Maximum daily photosynthesis was between 500 and 900 mg C/m^2/day.

12. Minimum oxygen 5 m above the lake bottom occurred in October of both years. In 1981 the minimum was 4.4 mg/liter and in 1982 it was 4.6 mg/liter. This represents about 50% loss of oxygen from the saturation values. The areal hypolimnetic oxygen deficit was 710 mg/m^2/day in 1981 and 630 mg/m^2/day in 1982.

13. Nutrient enrichment studies showed that phosphorus limitation prevails between 1 January and 15 July. After 15 July, the phytoplankton community is strongly nitrogen limited until September, when thickening of the epilimnion replenishes nitrogen supplies. At the beginning of October, strong phosphorus limitation is reestablished. Nutrient limitation is minimal during the month of May and during the last 3 weeks of September. Nitrogen and phosphorus limitations are rather closely balanced, as shown by the switch from one to the other over the seasonal cycle.

14. Studies of horizontal spatial variation over the lake show no statistical evidence of sustained differences between different parts of the lake. Randomly changing patterns of variation between stations are detectable but small for both biological and chemical variables. The lake can be treated as a functional unit.

15. A variety of trophic indicators indicate that Lake Dillon is mesotrophic.

16. Volume of water entering the lake by way of the three major rivers was almost twice as high in 1982 as in 1981. The total weighted concentration of both nitrogen and phosphorus was somewhat higher in 1982 than in 1981. The annual discharge-weighted averages for total phosphorus concentration in inflowing rivers varied between 6 and 20 μg/liter. Annual discharge-weighted nitrogen concentrations varied between 200 and 600 μg/liter in the three rivers. The average total phosphorus concentrations in all tertiary wastewater treatment plant effluents were raised considerably by short-term shutdowns and malfunctions.

17. The total amount of phosphorus entering the lake was 2900 kg in 1981 and 4800 kg in 1982. The comparable figures for nitrogen were 85,600 kg in 1981 and 151,400 kg in 1982. Expressed per unit lake area, the phosphorus loading corresponded to 0.29 g/m^2/year in 1981 and 0.42 g/m^2/year in 1982.

18. The total nutrient loading was divided according to sources on the basis of a detailed study of small representative watersheds and watershed segments. Equations for the following sources were developed which relate the amount of nutrient yield from a given land surface to the amount of runoff: background yield (undisturbed), nonpoint yield from residential area on sewer, nonpoint yield from urban area on sewer, yield from residential area on septic systems, and yield from interstate highway. Equations were also developed for the yield from the following sources that did not show dependence on the amount of runoff: ski slopes, mining, secondary treatment plants (package plants), tertiary treatment plants.

19. It was shown from an analysis of the stream segments that river valley bottoms incorporating wetland, standing water, and gravel beds accumulate significant fractions of the nutrient yield in dry years and export accumulated yield in wet years. An equation was developed to account for this storage and purging effect.

20. The yield equations and the storage equations were tested on the total nutrient yield in runoff for 1981 and 1982. The predicted runoff yield to the lake was 2800 kg and the observed was 2900 kg in 1981. In 1982, the predicted was 4600 and the observed was 4800. Nitrogen predictions were similarly close.

21. By use of the yield equations and information on nonrunoff nutrient yield, a complete breakdown was made of the phosphorus and nitrogen sources for Dillon in

1981 and 1982. In 1982, which is the more typical of a run of years under present trophic conditions, tertiary plants accounted for 15% of phosphorus loading, secondary plants for 2.1%, Climax Molybdenum mining for 2.2%, background runoff for 45.6%, precipitation for 13.1%, groundwater for 1.9%, dispersed nonpoint sources for 14.2%, and a major construction project in the Snake River bottom for 6.4%.

22. At the present time, total loading of the lake can be divided approximately into four quarters. The first two of these quarters are taken up by natural sources. The third quarter is taken up by sources that can be traced to human waste (septic fields and wastewater treatment plants), and the fourth quarter is accounted for by human activities that cannot be traced to waste disposal.

23. A model was developed with the following components: (a) a land use component that accepts information on land use and amount of runoff, (b) a trophic status component that accepts output from the land-use component, and (c) an effects component that accepts output from the trophic status component. This model predicts the nutrient yields by source and by watershed segments given any reasonable combination of future land uses. It also predicts the total loading and trophic status of the lake in terms of total phosphorus and chlorophyll *a*, and translates this trophic status prediction into economically significant measures such as transparency and hypolimnetic oxygen deficit. The model performs well on 1981 and 1982 data and is used to make predictions about five different development scenarios.

References

Ahlgren, I. 1979. Lake metabolism studies and results at the Institute of Limnology in Uppsala. *Arch. Hydrobiol. Beih.* 13:10–30.

American Public Health Assn. 1976. *Standard Methods for the Examination of Water and Wastewater.* Washington, D.C.: American Public Health Assn.

Armitage, K. B., and J. C. Tash. 1967. The life cycle of *Cyclops bicuspidatus thomasi.* S.A. Forbes in Leavenworth County State Lake, Kansas, U.S.A. (Copepoda). *Crustaceana* 13:94–102.

Bailey-Watts, A. E. 1978. A nine year study of the phytoplankton of the eutrophic and non-stratifying Loch Leven (Kinross, Scotland). *J. Ecol.* 66:741–771.

Bannister, T. T. 1974a. Production equations in terms of chlorophyll concentration, quantum yield, and upper limit to production. *Limnol. Oceanogr.* 19:1–12.

———. 1974b. A general theory of steady state phytoplankton growth in a nutrient saturated mixed layer. *Limnol. Oceanogr.* 19:13–30.

Bendschneider, K., and R. J. Robinson. 1952. A new spectrophotometric method for the determination of nitrate in sea water. *J. Mar. Res.* 11:87–96.

Bormann, F. H., and G. E. Likens. 1979. *Pattern and Process in a Forested Ecosystem.* New York: Springer-Verlag.

———, G. E. Likens, and J. S. Eaton. 1969. Biotic regulation of particulate and solution losses from a forest ecosystem. *BioScience* 19:600–610.

Bottrell, H. H., A. Duncan, C. M. Gliwicz, E. Grygierek, A. Herzig, A. Hilbricht-Ilkowska, H. Kurasawa, P. Larsson, and T. Weglenska. 1976. A review of some problems in zooplankton production studies. *Norwegian J. Zool.* 24:419–456.

Brunskill, G. J., D. Povoledo, B. W. Graham, and M. P. Stainton. 1971. Chemistry of surface sediments of sixteen lakes in the Experimental Lakes Area, northwestern Ontario. *J. Fish. Res. Bd. Canada.* 28:277–294.

Carlson, R. E. 1977. A trophic state index for lakes. *Limnol. Oceanogr.* 22:361–368.

Chapra, K. H., and S. C. Reckhow. 1983. *Engineering Approaches for Lake Management. Volume 1: Data Analysis and Empirical Modeling.* Boston: Butterworth.
Cole, G. A. 1955. An ecological study of the microbenthic fauna of two Minnesota lakes. *Amer. Midl. Nat.* 53:213–230.
Cornett, J. R., and F. H. Rigler. 1980. Areal hypolimnetic oxygen deficit: An empirical test of the model. *Limnol. Oceanogr.* 25:672–679.
Dillon, P. J., and F. H. Rigler. 1974. The phosphorus-chlorophyll relationship in lakes. *Limnol. Oceanogr.* 19:767–773.
Droop, N. R. 1973. Some thoughts on nutrient limitation in algae. *J. Phycol.* 9:264–272.
Edmondson, W. T. 1972. Nutrients and phytoplankton in Lake Washington. In *Nutrients and Eutrophication,* G. E. Likens, ed. Vol. 1, pp. 172–188. American Society of Limnology and Oceanography Special Symposia.
Eppley, R. W. 1981. Relations between nutrient assimilation and growth in phytoplankton with a brief review of estimates of growth rate in the ocean. In: *Physiological Bases of Phytoplankton Ecology,* T. Platt, ed. pp. 251–263. *Can. Bull. Fish. Aquat. Sci. Bull.* 210.
Fee, E. J. 1976. The vertical and seasonal distribution of chlorophyll in lakes of the Experimental Lakes Area, northwestern Ontario: Implications for primary production estimates. *Limnol. Oceanogr.* 21:767–783.
Fogg, G. E. 1975. *Algal Cultures and Phytoplankton Ecology.* Madison: Univ. of Wisconsin Press.
———, W. D. P. Stewart, P. Fay, and A. E. Walsby. 1973. *The Bluegreen Algae.* New York: Academic Press.
Ganf, G. G. 1975. Photosynthetic production and irradiance-photosynthesis relationships of the photoplankton from a shallow equatorial lake (Lake George, Uganda). *Oecologia* 18:165–183.
Gerhart, D. Z., and G. E. Likens. 1975. Enrichment experiments for determining nutrient limitation: Four methods compared. *Limnol. Oceanogr.* 20:649–653.
Gibson, C. E. 1978. Field and laboratory observations on the temporal and spatial variation of carbohydrate content in planktonic blue-green algae in Lough Neagh, Northern Ireland. *J. Ecol.* 66:97–115.
Glooschenko, W. A., J. E. Moore, and R. A. Vollenweider. 1973. Chlorophyll a distribution in Lake Huron and its relationship to primary productivity. *Great Lakes Res.* 16:40–49.
Goldman, C. R. 1972. The role of minor nutrients in limiting the productivity of aquatic ecosystems. *Am. Soc. Limnol. Oceanogr. Spec. Symp.* 1:21–38.
———. 1974. *Eutrophication of Lake Tahoe Emphasizing Water Quality.* EPA Report 660/3-74-034.
Golterman, H. L. 1975. *Physiological Limnology.* Amsterdam: Elsevier Scientific.
———. (ed.). 1976. *Interactions between Sediments and Fresh Water.* The Hague: Junk.
Grant, M. C., and W. M. Lewis, Jr. 1982. Precipitation chemistry in the Colorado Rockies. *Tellus* 34:74–88.
Grasshoff, K. 1976. *Methods of Seawater Analysis.* Weinheim: Verlag Chemie.
Gulati, R. D., K. Siewertsen, and G. Postema. 1982. The zooplankton: Its community structure, food and feeding, and role in the ecosystem of Lake Vechten. *Hydrobiologia* 95:127–163.
Harris, G. P. 1978. Photosynthesis, productivity, and growth: The physiological ecology of phytoplankton. *Arch. Hydrobiol. Beih. Ergebn. Limnol.* 10:1–171.
———. 1980. Temporal and spatial scales in phytoplankton ecology. *Can. J. Fish. Aquat. Sci.* 37:877–900.
Healey, F. P. 1979. Short-term responses of nutrient deficient algae to nutrient addition. *J. Phycol.* 15:289–299.

Heaney, S. I., and J. F. Talling. 1980. Dynamic aspects of dinoflagellate distribution patterns in a small productive lake. *J. Ecol.* 68:75–94.

Higgins, J. M., and B. R. Kim. 1981. Phosphorus retention models for Tennessee Valley Authority reservoirs. *Water Resour. Res.* 17:571–576.

Hobbie, J. E., and J. L. Tiwari. 1977. Ecosystem models vs. biological reality: Experiences in the systems analysis of an arctic pond. *Verh. Internat. Verein. Limnol.* 20:105–109.

Hutchinson, G. E. 1938. On the relation between oxygen deficit and the productivity and typology of lakes. *Int. Rev. ges. Hydrobiol. Hydrogr.* 36:336–355.

Hutchinson, G. E. 1957. *A Treatise on Limnology. Vol. 1. Geography, Physics, and Chemistry.* New York: Wiley.

Hutchinson, G. E. 1967. *A Treatise on Limnology. Vol. 2. Introduction to Lake Biology and the Limnoplankton.* New York: Wiley.

Hynes, H. B. M. 1970. *The Ecology of Running Waters.* Toronto: Univ. of Toronto Press.

Jones, A. R., and G. F. Lee. 1982. Recent advances in assessing impact of phosphorus loads on eutrophic-related water quality. *Water Res.* 16:503–515.

Jones, J. R., and R. W. Bachmann. 1976. Prediction of phosphorus and chlorophyll levels in lakes. *J. Water. Pollut. Cont. Fed.* 48:2176–2182.

Kalff, J., and R. Knoechel. 1978. Phytoplankton and their dynamics in oligotrophic and eutrophic lakes. *Ann. Rev. Ecol. Syst.* 9:475–495.

Kirk, J. T. O. 1975. A theoretical analysis of the contribution of algal cells to the alteration of light within natural waters. *New Phytol.* 75:11–20.

Komarek, J. 1976. Taxonomic review of the genera *Synechocystis* Sauv. 1982, *Synechococcus* Näg 1949, and *Cyanothece* gen. nov. (Cyanophyceae). *Arch. Protistenk.* 118:119–179.

Lambou, V. W., S. C. Hern, W. D. Taylor, and L. R. Williams. 1982. Chlorophyll, phosphorus, secchi disk and trophic state. *Water Resources Bull.* 18:807–813.

Larsen, D. P., and H. T. Mercier. 1976. Phosphorus retention capacity of lakes. *J. Fish. Res. Bd. Canada* 33:1742–1750.

Lasenby, D. C. 1975. Development of oxygen deficits in 14 southern Ontario lakes. *Limnol. Oceanogr.* 20:993–999.

Lean, D. R. S., and F. R. Pick. 1981. Photosynthetic response of lake plankton to nutrient enrichment: A test of nutrient limitation. *Limnol. Oceanogr.* 26:1001–1019.

LeBlanc, A., A. Maire, and A. Aubin. 1981. Ecologie et dynamique des populations de Copepodes (Cyclopoida) des principaux types de milieux astatiques temporaires de la zone temperee du Quebec Meridional. *Can. J. Zool.* 59:722–732.

Legnerova, J. 1969. The systematics and ontogenesis of the genera *Ankistrodesmus* Corda and *Monoraphidium* gen. nov. In *Studies in Phycology,* B. Fott, ed. pp. 75–122. Prague: Czechoslovak Acad. Sci.

Leopold, L. B., M. G. Wolman, and J. P. Miller. 1964. *Fluvial Processes in Geomorphology.* San Francisco: Freeman.

Levine, S. N., and D. W. Schindler. 1980. Radiochemical analysis of orthophosphate concentrations and seasonal changes in the flux of orthophosphate to seston in Canadian Shield lakes. *Can. J. Fish. Aquat. Sci.* 37:479–487.

Lewis, W. M., Jr. 1974. Primary production in the plankton community of a tropical lake. *Ecol. Monogr.* 44:377–409.

———. 1979. Spatial distribution of the phytoplankton in a tropical lake. *Int. Rev. ges. Hydrobiol.* 63:619–635.

———. 1980. Comparison of spatial and temporal variations in the zooplankton of a lake by means of variance components. *Ecology* 59:666–671.

———. 1983a. A revised classification of lakes based on mixing. *J. Fish. Aquat. Sci.* 40:1779–1787.

——. 1983b. Interception of atmospheric fixed nitrogen: An explanation of scum formation in nutrient-stressed blue-green algae. *J. Phycol.* 19:534–536.

——, and D. Canfield. 1977. Dissolved organic carbon in some dark Venezuelan waters and a revised equation for spectrophotometric determination of dissolved organic carbon. *Arch. Hydrobiol.* 79:441–445.

——, and M. C. Grant. 1978. Sampling and chemical interpretations of precipitation for mass balance studies. *Water Resources Res.* 14:1098–1104.

——, and M. C. Grant. 1979. Relationships between stream discharge and yield of dissolved substances from a Colorado mountain watershed. *Soil Sci.* 128:253–363.

——, and M. C. Grant. 1980a. Change in the output of ions from a watershed as a result of acidification of precipitation. *Ecology* 60:1093–1097.

——, and M. C. Grant. 1980b. Relationship between snow cover and winter losses of dissolved substances from a mountain watershed. *Arct. Alp. Res.* 12(1):11–17.

——, and J. F. Saunders, III. 1979. Two new integrating samplers for zooplankton, phytoplankton, and water chemistry. *Arch. Hydrobiol.* 85:245–249.

——, and F. H. Weibezahn. 1976. Chemistry, energy flow, and community structure in some Venezuelan fresh waters. *Arch. Hydrobiol.* Suppl. 50(2/3):145–207.

Likens, G. E. 1975. Primary production of inland aquatic ecosystems. In *The Primary Productivity of the Biosphere,* H. Lieth and R. H. Whittaker, eds. pp. 185–202. New York: Springer-Verlag.

——, F. H. Bormann, R. S. Pierce, J. S. Eaton, and N. M. Johnson. 1977. *Biogeochemistry of a Forested Ecosystem.* New York: Springer-Verlag.

Lund, J. W. G. 1961. The periodicity of μ-algae in three English lakes. *Verh. Internat. Verein. Limnol.* 14:147–154.

——. 1965. The ecology of freshwater phytoplankton. *Biol. Rev.* 40:231–293.

——. 1970. Primary Production. *Proc. Soc. Wat. Treat. Exam.* 19:332–358.

——. 1971. The seasonal periodicity of three planktonic desmids in Windermere. *Mitt. Internat. Verein. Limnol.* 19:3–25.

——. 1979. Changes in the phytoplankton of an English lake, 1945–1977. *Hydrobiol. J.* 14:6–21.

——. 1981. Investigations on phytoplankton with special reference to water usage. *Freshwat. Biol. Assoc. Occasional Publ.* 13:1–64.

Makarewicz, J. C., and G. E. Likens. 1975. Niche analysis of a zooplankton community. *Science* 190:1000–1003.

——, and G. E. Likens. 1979. Structure and function of the zooplankton community of Mirror Lake, New Hampshire. *Ecological Monogr.* 49:109–127.

Maloney, T. E., W. E. Miller, and T. Shiroyama. 1972. Algal responses to nutrient additions in natural waters. I. Laboratory assays. *Am. Soc. Limnol. Oceanogr.* Special Symposia, Vol. 1.

Margalef, R. 1978. Life-forms of phytoplankton as survival alternatives in an unstable environment. *Oceanol. Acta* 1:493–509.

Marker, A. F. H., C. A. Crowther, and R. J. M. Gunn. 1980. Methanol and acetone as solvents for estimating chlorophyll and phaeopigments by spectrophotometry. *Arch. Hydrobiol. Beih.* 14:52–69.

McQueen, D. J. 1969. Reduction of zooplankton standing stocks by predaceous *Cyclops bicuspidatus thomasi* in Marion Lake, British Columbia. *J. Fish. Res. Bd. Can.* 26:1605–1618.

Moll, R. A., and E. F. Stoermer. 1982. A hypothesis relating trophic status and subsurface chlorophyll maxima of lakes. *Arch. Hydrobiol.* 94(4):425–440.

Mortimer, C. H. 1941. The exchange of dissolved substances between mud and water. *J. Ecol.* 29:280–329.

Murphy, J., and J. P. Riley. 1962. A modified single solution method for the determination of phosphate in natural waters. *Anal. Chim. Acta* 27:31–36.

Moss, B. 1972. Studies on Gull Lake, Michigan. Seasonal and depth distribution of phytoplankton. *Freshwater Biol.* 2:289–307.

Nalewajko, C., and D. R. S. Lean. 1980. Phosphorus. In *The Physiological Ecology of Phytoplankton,* I. Morris, ed. pp. 235–258. Berkeley: Univ. Calif. Press.

Nauwerck, A. 1963. Die Beziehungen zwischen Zooplankton und Phytoplankton im See Erken. *Symb. Bot. Upsal.* 17:1–163.

Neill, W. E., and A. Peacock. 1980. Breaking the bottleneck: Interactions of invertebrate predators and nutrients in oligotrophic lakes. In *Evolution and Ecology of Zooplankton Communities,* W. C. Kerfoot, ed. pp. 715–724. Hanover: The Univ. Press of New England.

Nelson, W. C. 1971. *Comparative Limnology of Colorado-Big Thompson Project Reservoirs and Lakes.* Colorado Division of Wildlife Report (mimeo).

———. 1981. Large lake and reservoir limnological studies. *Col. Fish. Res. Rev.* DOW-R-R-F78-80:6–8.

O.E.C.D. 1982. Eutrophication of Waters. Paris: O.E.C.D.

Park, A. R., T. W. Groden, and C. J. Desormeau. Modifications to the model CLEANER requiring further research. In *Perspectives on Lake Ecosystem Modeling,* D. Scavia and A. Robertson, eds. pp. 87–108. Ann Arbor: Ann Arbor Science.

Parsons, T. R., M. Takahashi, and B. Hargrave. 1977. *Biological Oceanographic Processes.* New York: Pergamon.

Pennak, R. W. 1949. Annual limnological cycles in some Colorado reservoir lakes. *Ecol. Monogr.* 19:233–267.

Platt, T., and C. Filion. 1973. Spatial variability of the productivity: Biomass ratio for phytoplankton in a small marine basin. *Limnol. Oceanogr.* 18:743–749.

Rai, H. (ed.). 1980. The measurement of photosynthetic pigments in freshwaters and standardization of methods. *Arch. Hydrobiol. Beih. Ergeb. Limnol.* 14:1–106.

Ragotzkie, R. A. 1978. Heat budgets of lakes. In *Lakes: Chemistry, Geology, Physics,* A. Lerman, ed. pp. 1–19. New York: Springer-Verlag.

Ramberg, L. 1976. Relations between phytoplankton and environment in two Swedish forest lakes. *Scripta Limnol. Upsal.* 426:1–97.

———. 1979. Relations between phytoplankton and light climate in two Swedish forest lakes. *Internat. Rev. ges. Hydrobiol.* 64:749–782.

Rhee, G. Y. 1980. Continuous culture in phytoplankton ecology. *Adv. Aquat. Microbiol.* 2:151–203.

Rieman, B. 1980. A note on the use of methanol as an extraction solvent for chlorophyll a determination. *Arch. Hydrobiol. Beih. Ergebn. Limnol.* 14:70–78.

Rigler, F. H. 1968. Further observations inconsistent with the hypothesis that molybdenum blue method measures orthophosphate in lake water. *Limnol. Oceanogr.* 13:7–13.

Rohlf, J. F., and R. Sokal. 1969. *Statistical Tables.* San Francisco: Freeman.

Saunders, G. W., F. B. Trama, and R. W. Bachmann. 1962. Evaluation of a modified C-14 technique for shipboard estimation of photosynthesis in large lakes. *Great Lakes Res. Div.* Publ. No. 8.

Sawyer, C. N. 1947. Fertilization of lakes by agriculture and urban drainage. *J. New England Water Works. Assoc.* 61:109–127.

Schindler, D. W., R. W. Newbury, K. G. Beaty, and P. Campbell. 1976. Natural water and chemical budgets for a small Precambrian lake basin in Central Canada. *J. Fish. Res. Bd. Canada* 33:2526–2543.

Smith, R. C. 1968. The optical characterization of natural waters by means of an "extinction coefficient." *Limnol. Oceanogr.* 13:423–429.

Smith, R. E., and R. J. Kernehan. 1981. Predation by the free-living copepod, *Cyclops bicuspidatus thomasi* on larvae of the striped bass and white perch. *Estuaries* 4: 81–83.

Smith, V. H. 1982. The nitrogen and phosphorus dependence of algal biomass in lakes: An empirical and theoretical analysis. *Limnol. Oceanogr.* 27:1101–1112.

Solórzano, L. 1969. Determination of ammonia in natural waters by the phenolhypochlorite method. *Limnol. Oceanogr.* 14:799–801.

———, and J. Sharp. 1980a. Determination of total dissolved phosphorus and particulate phosphorus in natural waters. *Limnol. Oceanogr.* 25:754–758.

———, and J. Sharp. 1980b. Determination of total dissolved nitrogen in natural waters. *Limnol. Oceanogr.* 25:751–754.

Strom, K. M. 1931. Fetovatn: A physiographic and biological study of a mountain lake. *Arch. Hydrobiol.* 22:491–536.

Stumm, W., and P. Baccini. 1978. Man-Made Chemical Perturbations of Lakes. In *Lakes–Chemistry, Geology, Physics,* A. Lerman, ed. pp. 91–123. New York: Springer-Verlag.

———, and J. J. Morgan. 1981. *Aquatic Chemistry.* New York: Wiley.

Taguchi, S. 1976. Relationship between photosynthesis and cell size of marine diatoms. *J. Phycol.* 12:185–189.

Talling, J. F. 1974. General outline of spectrophotometric methods. In *A Manual on Methods of Measuring Primary Production in Aquatic Environments,* R. A. Vollenweider, ed. pp. 22–26. London: Blackwell.

———. 1971. The underwater light climate as a controlling factor in the production ecology of freshwater phytoplankton. *Mitt. Internat. Verein. Limnol.* 19:214–243.

———. 1976. The depletion of carbon dioxide from lake water by phytoplankton. *J. Ecol.* 64:79–121.

———. 1979. Factor interactions and implications for the prediction of lake metabolism. *Arch. Hydrobiol. Beih.* 13:96–109.

———. 1982. Utilization of solar radiation by phytoplankton. In *Trends in Photobiology,* C. Helene, M. Charlier Th. Montenay-Garestier, and G. Laustriat, eds. pp. 619–631. New York: Plenum.

Utermöhl, H. 1958. Zur Vervollkommnung der quantitativen phytoplankton–Methodik. *Internat. Verein. Theor. Angew. Limnol. Mitt.* 9:1–38.

Vollenweider, R. A. 1968. Scientific fundamentals of the eutrophication of lakes and flowing waters, with particular reference to nitrogen and phosphorus as factors in eutrophication. *Organization for Economic Cooperation and Development Report.* DAS/CSI/68.27.

———. 1969. Possibilities and limits of the elementary models concerning the budget of substance in lakes. *Arch. Hydrobiol.* 66:1–36.

———, W. Rast, and J. Kerekes. 1980. The phosphorus loading concept and Great Lakes eutrophication. In *Phosphorus Management Strategies for Lakes,* R. C. Loehr, C. S. Martin, and W. Rast, eds. pp. 207–234. Ann Arbor: Ann Arbor Science.

Vollenweider, R. A. 1965. Calculation models of photosynthesis-depth curves and some implications regarding day rate estimates of primary production measurements. *Mem. Ist. Ital. Idrobiol.* Suppl. 18:425–457.

Wahlstrom, E. E., and D. Q. Hornback. 1962. Geology of the H. D. Roberts Tunnel, Colorado. *Geol. Soc. Am. Bull.* 73:1477–1498.

Welch, E. B. 1980. *Ecological Effects of Wastewater.* New York: Cambridge University Press.

Wood, E. D., F. A. J. Armstrong, and F. A. Richards. 1967. Determination of nitrate in sea water by cadmium-copper reduction to nitrite. *J. Mar. Biol. Assoc. U.K.* 47: 23–31.

Wolk, C. P. 1973. Physiology and cytological chemistry of blue-green algae. *Bacteriol. Rev.* 37:32–101.

Wright, R. F. 1974. Forest Fire: Impact on the hydrobiology, chemistry and sediments of small lakes in northeastern Minnesota. Interim Report. No. 10, Limnology Research Center. Minneapolis: Univ. of Minnesota.

Index

Ecological Studies

Volume 38
The Ecology of a Salt Marsh
By L. Pomeroy and R. G. Wiegert
1981. xiv, 271p. 57 figures. cloth
ISBN 0-387-90555-3

Volume 39
Resource Use by Chaparral and Matorral
A Comparison of Vegetation Function in Two Mediterranean Type Ecosystems
By P. C. Miller
1981. xviii, 455p. 118 figures. cloth
ISBN 0-387-90556-1

Volume 40
Ibexes in an African Environment
Ecology and Social System of the Walia Ibex in the Simen Mountains, Ethiopia
By B. Nievergelt
1981. ix, 189p. 40 figures. cloth
ISBN 3-540-10592-1

Volume 41
Forest Island Dynamics in Man-Dominated Landscapes
Edited by R. L. Burgess and D. M. Sharpe
1981. xviii, 310p. 61 figures. cloth
ISBN 0-387-90584-7

Volume 42
Ecology of Tropical Savannas
Edited by B. J. Huntley and B. H. Walker
1982. xi, 669p. 262 figures. cloth
ISBN 3-540-11885-3

Volume 43
Mediterranean-Type Ecosystems
The Role of Nutrients
Edited by F. J. Kruger, D. T. Mitchell, and J. U. M. Jarvis
1983. xiii, 552p. 143 figures. cloth
ISBN 3-540-12158-7

Volume 44
Disturbance and Ecosystems
Components of Response
Edited by H. A. Mooney and M. Godron
1983. xvi, 294p. 82 figures. cloth
ISBN 3-540-12454-3

Volume 45
The Effects of SO_2 on a Grassland
A Case Study in the Northern Great Plains of the United States
Edited by W. K. Lauenroth and Eric M. Preston
1984. approx. 224p. 86 figures. cloth
ISBN 0-387-90943-5

Volume 46
Eutrophication and Land Use
Lake Dillon, Colorado
By William M. Lewis, Jr., James F. Saunders, III, David W. Crumpacker, and Charles Brendecke
1984. approx. 224p. 67 figures. cloth
ISBN 0-387-90961-3

Volume 47
Pollutants in Porous Media
Edited by B. Yaron, J. Goldshmid, and G. Dagan
1984. approx. 350p. approx. 116 figures. cloth
ISBN 0-540-13179-5